THE
NEW MATH
MADE SIMPLE

by

ALBERT F. KEMPF

MADE SIMPLE BOOKS

DOUBLEDAY & COMPANY, INC.

GARDEN CITY, NEW YORK

Library of Congress Catalog Card Number 66–12224
Copyright © 1966 by Doubleday & Company, Inc.
All Rights Reserved
Printed in the United States of America

TABLE OF CONTENTS

CHAPTER ONE

SETS, NUMBERS, NUMERATION

CHAPTER TWO

ADDITION AND SUBTRACTION OF WHOLE NUMBERS

CHAPTER THREE

MULTIPLICATION AND DIVISION OF WHOLE NUMBERS

CHAPTER FOUR

THE SET OF INTEGERS

CHAPTER FIVE

SOLVING EQUATIONS AND PROBLEMS

CHAPTER SIX

RATIONAL NUMBERS

CHAPTER SEVEN

SETS OF POINTS

CHAPTER EIGHT

MEASURE AND MEASUREMENT

CHAPTER NINE

RATIO, PROPORTION, PER CENT

CHAPTER TEN

CONGRUENCE AND SIMILARITY

CHAPTER ELEVEN

PERIMETER, AREA, VOLUME

CHAPTER TWELVE

PROBABILITY AND STATISTICS

INTRODUCTION

For many years the major emphasis in elementary school mathematics has been on the mechanical aspects of computation. This has created the erroneous and misleading idea that this is all there is to mathematics.

You can become very skillful in computation (learning to add, subtract, multiply, and divide) without really understanding why these mathematical processes work. With the advent of high-speed computers and desk calculators, the ability to compute is fast becoming an unsalable skill.

The tremendous advances taking place in mathematics and science demand that today's children must be taught the *why* as well as the *how* of mathematics. Today's society and, even more so, future societies will face problems that cannot even be predicted today. These problems will not be solved by rote-learned facts alone, but by the ability to think mathematically and to use mathematical methods of attacking the problems. In fact, these new problems will undoubtedly involve and require more new and as yet unknown mathematics.

This book is designed to help you go beyond the routine computational skills—to understand the basic structure of and oganization of elementary mathematical systems. In most cases, simple illustrations from the physical world are used to help you easily understand the mathematical ideas and concepts.

An alphabetized glossary of important mathematical terms is provided for your convenience. Answers to the exercises are included at the end of the book so that you can check your work. The exercises are intended to be used as a self-evaluation of understanding and not to be worked merely for the sake of obtaining the correct answer.

Your study of this book will be interesting and rewarding if you accept the attitude "Why does it work?" rather than "How does it work?"

THE NEW MATH MADE SIMPLE

CHAPTER ONE

SETS, NUMBERS, NUMERATION

SETS

In mathematics we are often concerned not with a single object, but with a collection of objects. For example, we hear about and speak of a collection of paintings, a row of chairs, or a set of dishes. Each of these collections is an example of a *set*.

A **set** is simply a collection of things considered as a single entity.

Definition 1–1:

The things contained in a given set are called **members** or **elements** of the set.

The members of a collection of paintings are the individual paintings in that set. The members of a row of chairs are the individual chairs in that row. The members of a set of dishes are the individual cups, saucers, plates, etc., in that set.

One method of naming sets is shown below.

$$A = \{\text{Bob, Bill, Tom}\}$$

This is read, "A is the set whose members are Bob, Bill, and Tom." Capital letters are usually used to denote sets. The braces, { }, merely denote a set. The names of the members of the set are listed, separated by commas, and then enclosed within braces.

An alternate use of the brace notation is illustrated below.

W = {Monday, Tuesday, Wednesday, Thursday, Friday, Saturday, Sunday}
W = {the days of the week}

The first of these examples lists or tabulates the members of the set W. In the second example a descriptive phrase is enclosed within braces. The latter example is read, "W is the set of days of the week."

Using the set W above, we can say:

Monday *is a member of* W.
Saturday *is a member of* W.

We can abbreviate the phrase "is a member of" by using the Greek letter epsilon, ϵ, to stand for this phrase. Then we can say:

Monday ϵ W
Saturday ϵ W

The slash line or slant bar, /, is often used to negate the meaning of a mathematical symbol. The mathematical symbol $\epsilon\!\!\!/$ is read, "is not a member of." For set W we can then say:

John $\epsilon\!\!\!/$ W (John is not a member of W.)
April $\epsilon\!\!\!/$ W (April is not a member of W.)

The symbols denoting the individual members of a set are generally lower-case letters of our alphabet, such as a, b, c, d, and so on.

Exercises 1–1:

Name the members of each of the following sets.

1. The set of the Great Lakes
2. The set of the last 3 months of the year
3. The set of states in the U.S. bordering the Gulf of Mexico
4. The set of men over 15 feet tall
5. The set of months in a year
6. The set of states in the U.S. whose names begin with the letter A

Write a description of each of the following sets.

7. $A = \{a,b,c,d\}$
8. $B = \{a,e,i,o,u\}$
9. $C = \{x,y,z\}$

Use the sets given in questions 7–9 above and insert the symbol ϵ or $\not\epsilon$ in each blank to make the following sentences true.

10. $a \_\_ A$ 13. $y \_\_ A$
11. $a \_\_ B$ 14. $y \_\_ B$
12. $a \_\_ C$ 15. $y \_\_ C$

THE EMPTY SET

Perhaps the preceding Exercise 4 (the set of men over 15 feet tall) caused you to wonder whether a set had been described. Although it seems natural to think of a set as having at least 2 numbers, it is mathematically convenient to consider a single object as a set (*a unit set*). It is also convenient to consider a collection containing no members as a set, called the *empty* set, or the null set, or the void set.

Definition 1–2:

The empty set is the set that contains no members.

The empty set is usually denoted by $\emptyset$ (a letter from the Scandinavian alphabet). $\emptyset$ is read, "the empty set." We can also indicate the absence of members by denoting the empty set by { }.

Other examples of the empty set are: the set of cookies in an empty cookie jar; the set of all living men over 200 years old; or the set of months in our year which contain more than 50 days.

SUBSETS

It is often necessary to think of sets that are "part of" another set or are "sets within a set." The set of chairs (C) in a room is a set within the set of all pieces of furniture (F) in that room. Obviously, every chair in the room is a member of set C and also a member of set F. This leads to the idea of a subset.

Definition 1–3:

"Set A is a **subset** of set B" means that every member of set A is also a member of set B. An equivalent definition of a subset might be:

Definition 1–4:

"Set A is a **subset** of set B" if set A contains no member that is not also in set B.

We can abbreviate the phrase "is a subset of" by using the conventional symbol $\subset$. $A \subset B$ means "set A is a subset of set B," or simply "A is a subset of B."

By using the symbolism already established, we can concisely state Definition 1–3 as follows:

$A \subset B$ if for every $x \epsilon A$ then $x \epsilon B$.
Consider the following sets.
$R = \{a,b,c,d,e\}$
$S = \{a,c,e\}$

Every member of set S is also a member of set R. Hence, $S \subset R$. R is not a subset of S $(R \not\subset S)$ because R contains members (b and d) which are not members of S.

All of the possible subsets of set S are given below.

$\{a\} \subset S$	$\{a,c\} \subset S$
$\{c\} \subset S$	$\{a,e\} \subset S$
$\{e\} \subset S$	$\{c,e\} \subset S$
$\{ \} \subset S$	$\{a,c,e\} \subset S$

The last two subsets of S, as listed above, can lead us to some general conclusions about the subset relation.

Is the empty set a subset of every set? By Definition 1–4, the empty set contains no member which is not also a member of any given set. Hence, we say that *the empty set is a subset of every set*.

Since $S = \{a,c,e\}$ and $\{a,c,e\} \subset S$, we are tempted to ask: "Is every set a subset of itself?" Regardless of the set we choose, every member of the set is obviously a member of the set. Hence, we say that *every set is a subset of itself*.

Exercises 1–2:

Consider the following sets. Then write the symbol $\subset$ or $\not\subset$ in each blank so that the following become true sentences.

$A = \{a,b,c,d,e\}$ $B = \{b,d,e,g\}$ $C = \{b,d\}$

1. $B \_\_ A$ 4. $C \_\_ B$
2. $B \_\_ C$ 5. $C \_\_ A$
3. $B \_\_ B$ 6. $\emptyset \_\_ C$

List all of the possible subsets of each of the following sets.

7. $D = \{x,y\}$
8. $E = \{a,b,c,d\}$

Compare the number of subsets and the number of members of set D, E, and the previously used set $S = \{a,c,e\}$.

9. Can you discover a formula for finding the number of subsets of any set?

SET EQUALITY

Consider the following sets.

$A = \{r,s,t,u\}$
$B = \{t,r,u,s\}$

Since each set contains identically the same members, we say that set A *is equal to* set B or simply $A = B$.

Definition 1–5:

If A and B are names for sets, $A = B$ means that set A has identically the same members as set B, or that A and B are two names for the same set.

Note that the order in which the members are named does not matter. For example, $\{a,b,c\} = \{c,a,b\} = \{b,a,c\}$.

Whenever the equal sign $(=)$ is used, as in $A = B$ or $1 + 2 = 3$, it means that the symbols on either side of it name precisely the same thing.

Consider the following sets.

$K = \{p,q,r,s\}$
$M = \{r,v,x,z\}$

Since K and M do not contain identically the same members we say K *is not equal to* M, or simply $K \neq M$.

Exercises 1–3:

Use the sets named below and write $=$ or $\neq$ in each blank so that true sentences result.

$A = \{1,2,3,4\}$
$B = \{a,e,i,o,u\}$
$C = \{\text{the first four counting numbers}\}$
$D = \{\text{the vowels in our alphabet}\}$
$E = \{3,2,1,4\}$
$F = \{o,i,a,w\}$

1. $A \_\_ B$
2. $A \_\_ C$
3. $A \_\_ D$
4. $A \_\_ E$
5. $A \_\_ F$
6. $B \_\_ C$
7. $B \_\_ D$
8. $B \_\_ E$
9. $B \_\_ F$
10. $E \_\_ F$

EQUIVALENT SETS

Suppose you had a set of cups and a set of saucers. Someone asks, "Are there more cups or more saucers?" Would you have to count the objects in each set to answer the question?

All you need do is place one cup on each saucer until all of the members of one of the sets have been used. If there are some cups left over, then there are more cups than saucers. If there are some saucers left over, then there are more saucers. In case each cup is paired with one and only one saucer and each saucer is paired with one and only one cup, we say the sets are matched one-to-one or that there is a one-to-one correspondence between the sets.

Definition 1–6:

There is a **one-to-one correspondence** between sets A and B if every member of A is paired with one member of B and every member of B is paired with one member of A.

The following illustration shows the six ways of establishing a one-to-one correspondence between the two sets.

The existence of a one-to-one correspondence between two sets has nothing to do with the way in which the pairing is done.

Definition 1–7:

Two sets are **equivalent** if there is a one-to-one correspondence between the two sets.

Note that the idea of equivalent sets is not the same as that of equal sets. That is, two sets are equal if they have identically the same members. Two equivalent sets may have different members just so there exists a one-to-one correspondence between them. For example:

{a,b,c,d} is equivalent to {r,s,t,u}.
{a,b,c,d} is not equal to {r,s,t,u}.
{a,b,c,d} is equal to {c,a,d,b}.
{a,b,c,d} is equivalent to {c,a,d,b}.

Exercises 1–4:

Draw matching lines to show a one-to-one correspondence between the sets in each pair.

1. {a,b,c,d}

   {w,x,y,z}
2. {1,2,3,4,5,6}

   {2,4,6,8,10,12}

NUMBERS

Let us consider the collection of all sets that are equivalent to {a,b,c}. For convenience, let us denote a set by drawing a ring around the collection of objects.

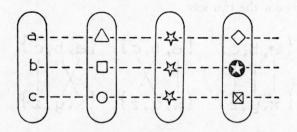

The only thing alike about all of these sets is that their members can be matched one-to-one. That is, they are equivalent sets. The thing that is alike about these sets is called the *number three*.

Of course, other sets belong to this collection also—the set of wheels on a tricycle, the set of people in a trio, and the set of sides of a triangle.

The number three has many names—III, 2 + 1, 3, and many more. Each of these names is called

a **numeral**. A *numeral* is a name for a number. The simplest numeral for the number three is 3.

With every collection of equivalent sets is associated a number, and with each number is associated a simplest numeral.

Set	Number	Simplest Numeral
{ }	zero	0
{a}	one	1
{c,d}	two	2
{x,y,z}	three	3
.	.	.
.	.	.
.	.	.

The dots indicate that we can extend each of the above columns. The set of numbers so derived is called the set of cardinal numbers or the set of *whole numbers*.

Since we usually begin counting "one, two, three, . . ." we call {1,2,3,4,5, . . .} the set of counting numbers or the set of *natural numbers*.

Set of whole numbers: {0,1,2,3,4, . . .}
Set of natural numbers: {1,2,3,4, . . .}

Exercises 1–5:

Write the simplest numeral for the number associated with each of the following sets.

1. {q,r,s,t,w,x,y,z}
2. {the days of the week}
3. {1,2,3,4,5,6,7,8,9}
4. {the months of the year}
5. {all three-dollar bills}
6. {John, James, Jean, Joe}
7. {Presidents of the U.S.}
8. {states in the U.S.}

BASE-TEN NUMERATION

Because of the random arrangement of the members of the set shown below, you may have a hard time determining quickly the number of members in the set.

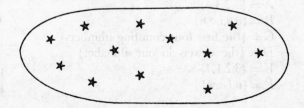

It is easier to determine the number of members if they are arranged as follows.

Since man has ten fingers, he probably matched members of a set one-to-one with his fingers and thereby grouped the objects as follows.

This led to his writing the symbol 12 to mean 1 set of ten and 2 more.

Finally it dawned on man that he could make any kind of grouping in his mind. Then he might group the members and name the number of members in any of the following ways.

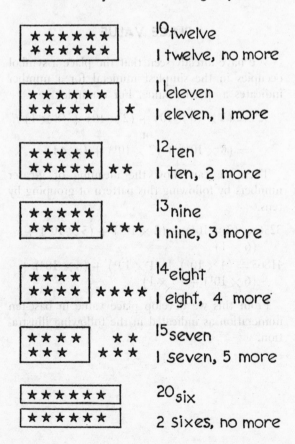

10_{twelve}

1 twelve, no more

11_{eleven}

1 eleven, 1 more

12_{ten}

1 ten, 2 more

13_{nine}

1 nine, 3 more

14_{eight}

1 eight, 4 more

15_{seven}

1 seven, 5 more

20_{six}

2 sixes, no more

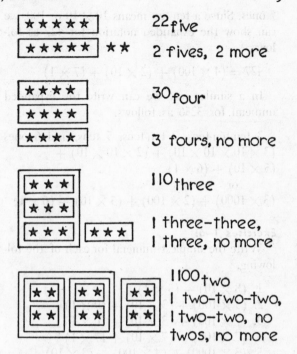

22_{five}

2 fives, 2 more

30_{four}

3 fours, no more

110_{three}

1 three-three, 1 three, no more

1100_{two}

1 two-two-two, 1 two-two, no twos, no more

Through the years man has found use for several of these methods of naming the number of the set. But his early ten-finger matching was deep-seated in his memory, and he most frequently grouped by tens. It is this grouping that leads to the *decimal* or *base-ten numeration* system.

Definition 1–8:

A *numeration system* is a planned scheme or way of naming numbers.

Let us agree that when a numeral is written without a number word to the lower right, such as 23, we shall mean base ten or grouping by tens. Then the numeral 23 means:

2 tens and 3 more
or
2 tens and 3 ones

Since 2 tens can be thought of as 2×10, and 3 ones can be thought of as 3×1, let us name 23 as follows:

$$23 = (2 \times 10) + (3 \times 1)$$

This is called the *expanded numeral* or the *expanded notation* for 23.

How can we name 427 in expanded notation? The numeral 427 means 4 ten-tens, 2 tens, and

7 ones. Since a ten-ten means 10×10 or 100, we can show the expanded notation for 427 as follows.

$$427 = (4 \times 100) + (2 \times 10) + (7 \times 1)$$

In a similar way we can write the expanded numeral for 3256 as follows.

3 ten-ten-tens, 2 ten-tens, 5 tens, and 6 ones
$$(3 \times 10 \times 10 \times 10) + (2 \times 10 \times 10) + (5 \times 10) + (6 \times 1)$$
or
$$(3 \times 1000) + (2 \times 100) + (5 \times 10) + (6 \times 1)$$

Exercises 1–6:

Write the simplest numeral for each of the following.

1. $(8 \times 10) + (5 \times 1)$
2. $(5 \times 100) + (3 \times 10) + (9 \times 1)$
3. $(7 \times 100) + (3 \times 10) + (0 \times 1)$
4. $(7 \times 100) + (7 \times 10) + (7 \times 1)$
5. $(3 \times 1000) + (4 \times 100) + (3 \times 10) + (2 \times 1)$
6. $(6 \times 1000) + (0 \times 100) + (5 \times 10) + (1 \times 1)$

Write the expanded numeral for each of the following.

7. 46	10.	82
8. 124	11.	3426
9. 629	12.	2041

EXPONENTS

It is inconvenient to write such things as $10 \times 10 \times 10$ and $5 \times 5 \times 5 \times 5$ whenever we express a number in expanded notation. Let us invent a short way of saying such things.

In $10 \times 10 \times 10$ we see that 10 is used 3 times in the multiplication. So let us write 10^3 to mean $10 \times 10 \times 10$.

Then $5 \times 5 \times 5 \times 5 = 5^4$ since 5 is used 4 times in the multiplication.

In 10^3, the number 10 is called the *base*, the number 3 is called the *exponent*, and the number named by 10^3 is called the *power*.

Base   *Exponent*
The number used   How many times
in the multiplication   the base is used

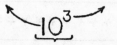

Power

Exercises 1–7:

Name each of the following as a power.

1. $10 \times 10 \times 10 \times 10 \times 10$
2. 10×10
3. $10 \times 10 \times 10 \times 10$
4. $7 \times 7 \times 7 \times 7$
5. $4 \times 4 \times 4 \times 4 \times 4 \times 4$

Write the meaning of each of the following.

6. 10^3	8. 10^7
7. 10^5	9. 6^4

PLACE VALUE

We have already seen that the place a symbol occupies in the simplest numeral for a number indicates a specific value. For example:

$$328 = (3 \times 10 \times 10) + (2 \times 10) + (8 \times 1)$$
$$\text{or}$$
$$= (3 \times 10^2) + (2 \times 10^1) + (8 \times 1)$$

Then we can show the meaning of greater numbers by following this pattern of grouping by tens.

$$3256 = (3 \times 10^3) + (2 \times 10^2) + (5 \times 10^1) + (6 \times 1)$$
$$41865 = (4 \times 10^4) + (1 \times 10^3) + (8 \times 10^2) + (6 \times 10^1) + (5 \times 1)$$

From this we develop place value in base-ten numeration as indicated in the following illustration.

APPROXIMATE NUMBERS

We often hear such remarks as "about 25,000 people" or "nearly 850 cars." These mean that the numerals do not name the exact number of objects, but only approximately that number. We often say that we "round off" to the nearest ten, the nearest hundred, the nearest thousand, and so on.

To express 7826 to the nearest hundred, think: Is 826 nearer to 800 or to 900?

Since it is nearer to 800, we replace 826 by 800. 7826 ≈ 7800 to the nearest hundred

We can't say 7826 = 7800, so we use the symbol ≈ to mean "is approximately equal to."

To express 52,946 to the nearest thousand, think:

Is 2946 nearer to 2000 or 3000?

Since it is nearer to 3000 replace 2946 by 3000. 52,946 ≈ 53,000 to the nearest thousand

To express 7250 to the nearest hundred, think: Is 250 nearer to 200 or 300?

In this case the number 250 is midway between 200 and 300. We must agree what to do in such a case. Let us agree to use whichever of 200 or 300 is even. That is, replace 250 by 200.

7250 ≈ 7200 to the nearest hundred

According to this agreement, 27,500 ≈ 28,000 to the nearest thousand since 7500 is midway between 7000 and 8000, and 8000 is even.

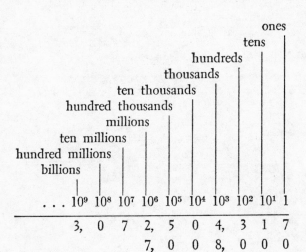

The commas are inserted merely to make it easy to read a numeral. They give no meaning whatsoever to the numeral.

The first numeral above is read: *three billion, seventy-two million, five hundred four thousand, three hundred seventeen.*

The second numeral above is read: *seven million, eight thousand.*

Exercises 1–8:

Write the simplest numeral for each of the following.

1. one billion, one hundred million, two thousand, eight hundred twenty-six
2. five million, one
3. seven hundred twelve thousand, three hundred nine
4. fifty-two million, eighteen

Exercises 1–9:

Express each of the following to the nearest thousand, then the nearest hundred, and then the nearest ten.

1. 28,562
2. 70,837
3. 53,149

CHAPTER TWO

ADDITION AND SUBTRACTION OF WHOLE NUMBERS

UNION OF SETS

We are accustomed to joining sets in our daily activities. For example, when you put some coins in your purse, you are joining two sets of coins—the set of coins already in your purse and the set of coins about to be put in your purse. This, and many more examples, form the basis for the idea of the *union* of two sets.

Definition 2–1:

The **union** of set A and set B, denoted by A ∪ B, is the set of all objects that are members of set A, of set B, or of both set A and set B. Consider the following sets.

$R = \{a,b,c,d\}$
$S = \{r,s,t\}$
$T = \{c,d,e,f\}$

According to the definition of union, we can form the following sets.

$R \cup S = \{a,b,c,d,r,s,t,\}$
$R \cup T = \{a,b,c,d,e,f\}$
$S \cup T = \{r,s,t,c,d,e,f\}$

For R ∪ T there is no need of repeating the names of members c and d. For example, suppose you are referring to a set of 3 girls—named Jane, Mary, and Pam. Then {Jane, Mary, Pam, Mary} is correct but not preferred since Mary is named twice and there are only 3 girls in the set.

Another way of illustrating sets and set operations is to use Venn diagrams. A Venn diagram is merely a closed figure used to denote the set of all points within the figure.

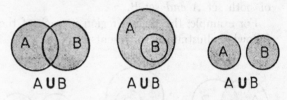

The shaded region in each of these illustrations indicates A ∪ B. Note that the union of two sets includes all of the members in both of the sets.

Exercises 2–1:

Use the following sets to form the union of each pair of sets given below.

$K = \{3,5,7,9\}$ $M = \{2,4,6,8\}$
$J = \{1,2,3\}$ $N = \{0,5,9\}$

1. $J \cup K$ 5. $J \cup N$
2. $K \cup M$ 6. $M \cup N$
3. $K \cup N$ 7. $N \cup M$
4. $J \cup M$ 8. $K \cup K$

INTERSECTION OF SETS

Suppose a teacher asked a class, "How many of you went to the game last night?" Then several children raised their hands. Those who raised their hands are members of the set of children in the class *and* they are also members of the set of all children who went to the game last night.

By using a Venn diagram we can illustrate this situation. Let $A = \{$all children in the class$\}$ and let $B = \{$all children who went to the game last night$\}$.

Then $C = \{$all children in A and also in B$\}$, and C is called the intersection of A and B.

Definition 2–2:

The **intersection** of set A and set B, denoted by $A \cap B$, is the set of all objects that are members of both set A *and* set B.

For example, the shaded region in each of the following illustrations represents $A \cap B$.

Consider the following sets.

$$X = \{g,h,i,j\}$$
$$Y = \{e,f,g,h\}$$
$$Z = \{a,b,c,d,e\}$$

Then: $\quad X \cap Y = \{g,h\}$
$$X \cap Z = \emptyset$$
$$Y \cap Z = \{e\}$$
$$X \cap X = \{g,h,i,j\}$$

Exercises 2–2:

Use the following sets to form the intersection of each pair of sets given below.

$$C = \{2,3,4,5,6\}$$
$$D = \{1,2,3,7,8\}$$
$$E = \{3,4,5,6\}$$
$$F = \{7,8,9,10\}$$

1. $C \cap D$ 5. $D \cap E$
2. $D \cap C$ 6. $D \cap F$
3. $C \cap E$ 7. $E \cap F$
4. $C \cap F$ 8. $E \cap E$

DISJOINT SETS

It is obvious that some sets have no members in common—such as $\{a,b,c\}$ and $\{r,s,t\}$.

Definition 2–3:

Set A and set B are called disjoint sets if they have no members in common. Or, set A and set B are *disjoint sets* if $A \cap B = \emptyset$.

In the following diagram, set A and set B do not intersect. Therefore, A and B are disjoint sets.

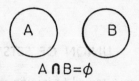

Consider the following sets.

$$R = \{f,g,h,j\}$$
$$S = \{a,b,c\}$$
$$T = \{a,h,j\}$$

Sets R and S have no members in common. Hence, R and S are disjoint sets.

Sets R and T are not disjoint sets since they both have h and j as members. Sets S and T are not disjoint sets since they both have a as a member.

Exercises 2–3:

Tell whether each statement below is *true* or *false*.

1. $\{q,r,s,t\}$ and $\{x,y,z\}$ are disjoint sets.
2. If $Q \cap R = \emptyset$, then Q and R are disjoint sets.
3. If $Q \subset R$, then Q and R are disjoint sets.
4. If $Q \subset R$ and $R \subset Q$, then $R = Q$.
5. If $R \cup Q = R$, then $Q \subset R$.
6. If $R \cap Q = R$, then $R \subset Q$.
7. If $R \cap Q = R \cup Q$, then $R = Q$.
8. If $C \cap D = \{5\}$, then $5 \epsilon C$ and $5 \epsilon D$.
9. If $C \cup D = \{3,4,5,6,7\}$, then $5 \epsilon C$ and $5 \epsilon D$.
10. If $x \epsilon H$, then $x \epsilon H \cup G$.

ADDITION

We already know that with each set there is associated a number. Let us use the symbol n(A) to mean "the number of set A." It is important to note that $n(A)$ is a name for a number.

For $A = \{a,b,c\}$, we have $n(A) = 3$.
For $B = \{g,h\}$, we have $n(B) = 2$.

Let us begin with two disjoint sets, C and D, and find $C \cup D$. That is, we will join set D to set C.

$$C = \{\square, \triangle, \star\} \qquad D = \{\bigcirc, \otimes\}$$

$$C \cup D = \{\square, \triangle, \star, \bigcirc, \otimes\}$$

$$n(C) = 3 \qquad\qquad n(D) = 2$$
$$n(C \cup D) = 5$$

From this illustration we can say what is meant by addition of whole numbers.

Definition 2–4:

For *disjoint sets* A and B, the **sum** of $n(A)$ and $n(B)$, denoted by $n(A) + n(B)$, is $n(A \cup B)$ or the number of the union set.

For the above illustration, we have $n(C) = 3$, $n(D) = 2$, and $n(C \cup D) = 5$.

$$n(C) + n(D) = n(C \cup D)$$
$$3 + 2 = 5$$

In an addition statement, the numbers being added are called **addends** and the resulting number is called the **sum**.

Caution! We *add numbers*, not sets. We write $3 + 2$, but we do not write $A + B$ for sets. We find the *union of sets*, not numbers. We write $A \cup B$, but we do not write $5 \cup 4$.

To find the sum of 6 and 4, we could think of disjoint sets A and B such that $n(A) = 6$ and $n(B) = 4$. Suitable sets might be:

$A = \{a,b,c,d,e,f\}$
$B = \{r,s,t,u\}$
$A \cup B = \{a,b,c,d,e,f,r,s,t,u\}$
$\quad n(A) + n(B) = n(A \cup B)$
$\quad\quad 6 + 4 = 10$

The numbers six and four are addends. The number ten is the sum. The numerals $6 + 4$ and 10 are two names for the sum. The numeral 10 is the simplest name for the number ten.

Exercises 2–4:
Find each sum.

1. $\begin{array}{r} 6 \\ +5 \\ \hline \end{array}$ $\begin{array}{r} 7 \\ +3 \\ \hline \end{array}$ $\begin{array}{r} 8 \\ +4 \\ \hline \end{array}$ $\begin{array}{r} 9 \\ +5 \\ \hline \end{array}$ $\begin{array}{r} 7 \\ +6 \\ \hline \end{array}$

2. $\begin{array}{r} 4 \\ +3 \\ \hline \end{array}$ $\begin{array}{r} 5 \\ +7 \\ \hline \end{array}$ $\begin{array}{r} 8 \\ +6 \\ \hline \end{array}$ $\begin{array}{r} 2 \\ +8 \\ \hline \end{array}$ $\begin{array}{r} 9 \\ +9 \\ \hline \end{array}$

3. $\begin{array}{r} 7 \\ +9 \\ \hline \end{array}$ $\begin{array}{r} 5 \\ +5 \\ \hline \end{array}$ $\begin{array}{r} 8 \\ +8 \\ \hline \end{array}$ $\begin{array}{r} 4 \\ +5 \\ \hline \end{array}$ $\begin{array}{r} 7 \\ +8 \\ \hline \end{array}$

4. $\begin{array}{r} 6 \\ +9 \\ \hline \end{array}$ $\begin{array}{r} 4 \\ +9 \\ \hline \end{array}$ $\begin{array}{r} 9 \\ +3 \\ \hline \end{array}$ $\begin{array}{r} 7 \\ +4 \\ \hline \end{array}$ $\begin{array}{r} 5 \\ +8 \\ \hline \end{array}$

5. $\begin{array}{r} 3 \\ +8 \\ \hline \end{array}$ $\begin{array}{r} 1 \\ +9 \\ \hline \end{array}$ $\begin{array}{r} 6 \\ +4 \\ \hline \end{array}$ $\begin{array}{r} 7 \\ +7 \\ \hline \end{array}$ $\begin{array}{r} 6 \\ +6 \\ \hline \end{array}$

ADDITION IS COMMUTATIVE

If you are to join two sets, you may wonder which set to join to which. Does the order of joining the sets change the union set? Let us examine such a situation.

$$A = \{\square, \bigcirc, \triangle\} \quad \overset{\text{join}}{\underset{\text{B to A}}{\curvearrowleft}} \quad B = \{a, b, c, d\}$$
$$A \cup B = \{\square, \bigcirc, \triangle, a, b, c, d\}$$

$$n(A) + n(B) = n(A \cup B)$$
$$3 + 4 = 7$$

Now let us reverse the order of joining the two sets.

$$A = \{\square, \bigcirc, \triangle\} \quad \overset{\text{join}}{\underset{\text{A to B}}{\curvearrowright}} \quad B = \{a, b, c, d\}$$
$$B \cup A = \{\square, \bigcirc, \triangle, a, b, c, d\}$$

$$n(B) + n(A) = n(B \cup A)$$
$$4 + 3 = 7$$

We notice that the union set is unchanged when the order of joining is reversed. We also note the following.

$$3+4=7 \text{ and } 4+3=7$$
$$\text{or}$$
$$3+4=4+3$$

The order of the addends can be changed but the sum remains the same. That is,

For all whole numbers a and b,
$$a+b=b+a.$$

We call this idea the **commutative property of addition.** Or we say that *addition is commutative*.

The phrase "for all whole numbers a and b" means that a and b can be replaced by numerals for any numbers in the set of whole numbers. They may be replaced by the same numeral or by different numerals. When a, b, or any other symbol is used in this manner, it is called a **placeholder** or a **variable** over a specified set of numbers.

Even though we may not know the sum of 557 and 3892, we know the following is true because addition is commutative.

$$557+3892=3892+557$$

Exercises 2–5:

Think of doing one activity of each pair given below and then doing the other. Do the following pairs illustrate a commutative property?

1. Put on your sock; put on your shoe
2. Take two steps forward; take two steps backward
3. Swim; eat
4. Write the letter "O"; then write the letter "N"
5. Go outside; close the door
6. Eat; brush your teeth

Complete each of the following sentences by using the commutative property of addition.

7. $3+7=7+\underline{}$
8. $\underline{}+15=15+8$
9. $36+\underline{}=17+36$
10. $156+13=\underline{}+156$
11. $129+47=\underline{}+\underline{}$
12. $\underline{}+\underline{}=218+326$
13. $327+\underline{}=56+\underline{}$
14. $651+87=\underline{}+\underline{}$

IDENTITY NUMBER OF ADDITION

Study the following unions of sets.

$$\{\ \} \cup \{a,b,c\} = \{a,b,c\}$$
$$\{a,b,c\} \cup \{\ \} = \{a,b,c\}$$

Notice that joining the empty set to a given set, or joining a given set to the empty set, does not change the given set.

The addition statements that correspond to the set operations above are:

$$0+3=3$$
$$3+0=3$$

Are the following sentences true?

$$7+0=7 \qquad 115+0=115 \qquad 721=0+721$$

Adding zero to any whole number b, or adding any whole number b to zero, leaves the number b unchanged.

Since zero is the only number with this special property, the number zero is called the **identity number of addition.**

For any whole number b,
$$0+b=b=b+0.$$

ADDITION IS ASSOCIATIVE

There are occasions when we join three sets. For example, we might combine a set of forks, a set of spoons, and a set of knives to form a set of silverware.

We might join the spoons to the forks, and then join the knives. Or we might join the knives to the spoons, and then join this set to the forks. Does the method of joining the sets change the resulting set?

Consider joining these sets.

$A=\{a,b,c\} \qquad B=\{g,h,j,k\} \qquad C=\{t,v\}$
$A \cup B = \{a,b,c,g,h,j,k\} \qquad B \cup C = \{g,h,j,k,t,v\}$
$$(A \cup B) \cup C = \{a,b,c,g,h,j,k,t,v\}$$
$$A \cup (B \cup C) = \{a,b,c,g,h,j,k,t,v\}$$

The ( ) in the last two statements indicate which two sets are joined first.

$(A \cup B) \cup C$ means to find $A \cup B$ first.
$A \cup (B \cup C)$ means to find $B \cup C$ first.

Joining sets makes us think of addition. We can add only two numbers at a time. How can we find the sum of three numbers, such as 3, 4, and 2?

Let us use the pattern established for joining three sets.

$$3 + 4 + 2 = (3 + 4) + 2$$
$$= 7 \quad + 2$$
$$= 9$$
$$3 + 4 + 2 = 3 + (4 + 2)$$
$$= 3 + \quad 6$$
$$= 9$$

The ( ) in $(3 + 4) + 2$ mean that 4 was added to 3 first. The ( ) in $3 + (4 + 2)$ mean that 2 was added to 4 first.

When finding the sum of three numbers we can group the first two addends or the last two addends and always get the same sum.

This idea is called the **associative property of addition**. Or we say that *addition is associative*. For all whole numbers *a*, *b*, and *c*,

$$(a + b) + c = a + (b + c).$$

We can add these first,

$$7 + 3 + 6$$

or we can add these first.

We can add these first, $\begin{array}{c} 5 \\ 8 \\ +2 \end{array}$ or add these first.

Notice that when we use the associative property of addition the order of the addends is *not* changed as it is when we use the commutative property of addition.

Exercises 2–6:

Three things are to be combined in each exercise below. Do not change their order, only the grouping. Do the combinations show an associative property?

1. Water, lemon juice, sugar
2. Sand, cement, water
3. Blue paint, red paint, green paint

Complete each of the following sentences by using the associative property of addition.

4. $5 + (7 + 6) = (5 + \underline{}) + \underline{}$
5. $17 + (15 + 32) = (\underline{} + \underline{}) + 32$
6. $(9 + 8) + 7 = \underline{} + (\underline{} + \underline{})$
7. $\underline{} + (\underline{} + \underline{}) = (13 + 12) + 6$
8. $(\underline{} + \underline{}) + \underline{} = 72 + (31 + 46)$

Find each sum below by using whichever grouping of addends makes the addition easier.

9. $7 + 3 + 6$ 11. $5 + 5 + 3$
10. $12 + 8 + 7$ 12. $9 + 13 + 7$

USING THE PROPERTIES OF ADDITION

We can show that $5 + (9 + 7) = 7 + (9 + 5)$ without using any addition facts.

$$5 + (9 + 7) = (5 + 9) + 7 \quad \text{Assoc. prop.}$$
$$= (9 + 5) + 7 \quad \text{Comm. prop.}$$
$$= 7 + (9 + 5) \quad \text{Comm. prop.}$$

Exercises 2–7:

Each of the following sentences is true because of the commutative property of addition, the associative property of addition, or both. Write the letter C, A, or both C and A to tell which property or properties are used.

1. $(9 + 8) + 3 = 9 + (8 + 3)$
2. $(9 + 8) + 3 = 3 + (9 + 8)$
3. $6 + (7 + 12) = 6 + (12 + 7)$
4. $6 + (7 + 12) = (6 + 12) + 7$
5. $(13 + 5) + 14 = 14 + (5 + 13)$
6. $(32 + 9) + 8 = 9 + (32 + 8)$
7. $13 + (9 + 7) = (13 + 9) + 7$
8. $13 + (9 + 7) = (9 + 7) + 13$
9. $13 + (9 + 7) = (13 + 7) + 9$
10. $a + (b + c) = (a + b) + c$

A NUMBER LINE

It is often helpful to think of a set of numbers as corresponding to points on a line. For the set of whole numbers, $\{0,1,2,3,4, \ldots\}$, we simply draw a line and locate two points labeled 0 and 1. The arrowheads at both ends of the picture indicate that the line extends indefinitely in both directions.

Then use the distance between the 0-point and the 1-point to locate points for 2, 3, 4, and so on. Such a drawing is called a **number line**.

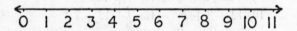

Of course, this line need not be drawn horizontal. It could just as well be in any other direction, as shown below. However, let us use the conventional horizontal arrangement.

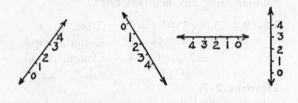

ADDITION ON A NUMBER LINE

We can picture addition by drawing arrows to represent the addends. For example, to represent $5 + 4 = \square$, we start by drawing an arrow from the 0-point to the 5-point to represent the addend 5. Then, starting at the head of this arrow (the 5-point) we draw another arrow extending in the same direction for 4 spaces (unit segments). The numeral for the sum is found directly below the head of the second arrow.

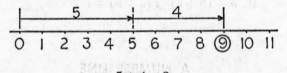

$$5 + 4 = 9$$

Using the conventional horizontal arrangement, addition is associated with "moving to the right" on a number line.

Exercises 2–8:

Write the addition sentence shown by each number-line drawing below.

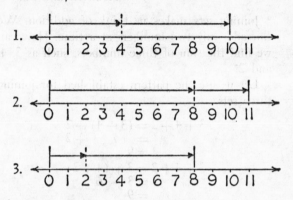

1.
2.
3.

ORDER OF WHOLE NUMBERS

If two sets are not equivalent, then one set contains more members than the other set. For example:

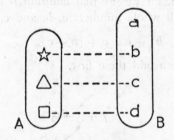

Set B has some members left unmatched after all of the members of set A have been matched. Set B has more members than set A, or set A has fewer members than set B. We use the symbol $<$ (read: *is less than*) and the symbol $>$ (read: *is greater than*) when comparing the numbers of two sets that are not equivalent.

$$n(A) < n(B) \text{ or } n(B) > n(A)$$
$$3 < 4 \text{ or } 4 > 3$$

The order of whole numbers is used when establishing a one-to-one correspondence between the whole numbers and selected points on a number line.

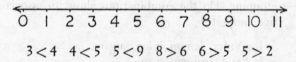

$$3 < 4 \quad 4 < 5 \quad 5 < 9 \quad 8 > 6 \quad 6 > 5 \quad 5 > 2$$

On a number line, the point farther to the right corresponds to the greater of two whole numbers.

THE SUM OF MORE THAN TWO ADDENDS

We can save time and effort by looking for sums of ten, sums of one hundred, and so on, when finding the sum of more than two addends. Think of finding the simplest numeral for the following sum.

$$3+6+5+4+7$$

We know that addition is associative, so we can use any grouping we please. We also know that addition is commutative, so we can change the order of addends as we please. By using these two properties of addition, we can think of the addition as follows:

$$3+6+5+4+7 = (3+6)+(5+4)+7$$
$$= (5+4)+(3+6)+7$$
$$= (5+4)+(6+3)+7$$
$$= 5+(4+6)+(3+7)$$
$$= 5+10+10$$
$$= 25$$

This type of thinking is used when we think about $3+6+5+4+7$ as follows:

$$3+6+5+4+7 =$$
$$10+10+5 = 25$$

Exercises 2–9:

Find each sum. Look for sums of ten or one hundred.

1.	5	8	13	25	97
	6	2	4	32	9
	+5	+7	+7	+75	+3

2.	7	4	24	19	37
	6	5	8	7	60
	4	5	2	1	13
	+3	+6	+6	+2	+40

THE ADDITION ALGORISM

Everyday problems make us aware of the need to have an easy method for operating with greater

numbers. For example, we may want to find the sum of 725 and 273. Both of these numbers have many names. We strive to name the numbers so that it is easy to find their sum.

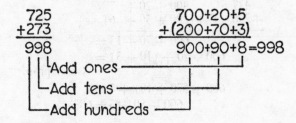

$$
\begin{array}{ll}
725 & 700+20+5 \\
+273 & +(200+70+3) \\
\hline
998 & 900+90+8=998
\end{array}
$$

Add ones
Add tens
Add hundreds

Another situation might require us to find the sum of 3528 and 4361.

$$
\begin{array}{ll}
3528 & 3000+500+20+8 \\
+4361 & +(4000+300+60+1) \\
\hline
7889 & 7000+800+80+9=7889
\end{array}
$$

This procedure, or algorism, of writing numerals and renaming numbers can be extended to finding the sum of greater numbers.

Exercises 2–10:

Find each sum.

1.	342	3751	35285
	+536	+4248	+24713

2.	235	5041	60027
	+542	+3806	+28951

3.	624	1826	1423
	+65	+153	+36504

4.	43	320	10726
	+325	+6468	+8070

RENAMING SUMS IN ADDITION

It may well be the case that the sum of the ones is greater than nine, or the sum of tens is greater than ninety, and so on. All we need do is rename such sums as shown in the following examples.

Rename the sum of the ones:

$$
\begin{array}{r} 427 \\ +256 \\ \hline \end{array}
\qquad
\begin{array}{r} 400 + 20 + 7 \\ +(200 + 50 + 6) \\ \hline 600 + 70 + 13 = \\ 600 + 70 + (10 + 3) = \end{array}
$$

Assoc. prop.

$$
600 + (70 + 10) + 3 =
$$
$$
600 + 80 + 3 = 683
$$

Rename the sum of the tens:

$$
\begin{array}{r} 3258 \\ +471 \\ \hline \end{array}
\qquad
\begin{array}{r} 3000 + 200 + 50 + 8 \\ +(400 + 70 + 1) \\ \hline 3000 + 600 + 120 + 9 = \\ 3000 + 600 + (100 + 20) + 9 = \end{array}
$$

Assoc. prop.

$$
3000 + (600 + 100) + 20 + 9 =
$$
$$
3000 + \quad 700 \quad + 20 + 9 = 3729
$$

Rename sums of tens and ones:

$$
\begin{array}{r} 3456 \\ +2378 \\ \hline \end{array}
\qquad
\begin{array}{r} 3000 + 400 + 50 + 6 \\ +(2000 + 300 + 70 + 8) \\ \hline 5000 + 700 + 120 + 14 = \end{array}
$$
$$
5000 + 700 + (100 + 20) + (10 + 4) =
$$

Assoc. prop.

$$
5000 + (700 + 100) + (20 + 10) + 4 =
$$
$$
5000 + \quad 800 \quad + \quad 30 \quad + 4 = 5834
$$

This procedure can be abbreviated by thinking about the addition as follows.

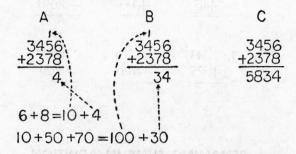

$$
6 + 8 = 10 + 4
$$
$$
10 + 50 + 70 = 100 + 30
$$

In A, $6 + 8 = 14$ and $14 = 10 + 4$. Write the 4 in ones place of the sum numeral and name the ten by writing a small reminder numeral 1 above the 5 in tens place of the first addend.

In B, $10 + 50 + 70 = 130$ and $130 = 100 + 30$. Write 3 in tens place of the sum numeral to name thirty; then write a reminder numeral 1 above the 4 in hundreds place in the first addend to name the hundred.

In C, the sum of hundreds is less than 1000 and the sum of the thousands is less than 10,000, so renaming is not needed.

The above procedure can be extended for addition of more than two numbers and for numbers whose numerals have a greater number of digits.

Exercises 2–11:
Find each sum.

1.
$$
\begin{array}{r} 3426 \\ +2595 \\ \hline \end{array}
\qquad
\begin{array}{r} 61897 \\ +17973 \\ \hline \end{array}
\qquad
\begin{array}{r} 567428 \\ +340754 \\ \hline \end{array}
$$

2.
$$
\begin{array}{r} 3058 \\ +4963 \\ \hline \end{array}
\qquad
\begin{array}{r} 47569 \\ +10753 \\ \hline \end{array}
\qquad
\begin{array}{r} 640596 \\ +365437 \\ \hline \end{array}
$$

3.
$$
\begin{array}{r} 3246 \\ 503 \\ +1174 \\ \hline \end{array}
\qquad
\begin{array}{r} 97654 \\ 7965 \\ +89348 \\ \hline \end{array}
\qquad
\begin{array}{r} 297254 \\ 34135 \\ +343048 \\ \hline \end{array}
$$

4.
$$
\begin{array}{r} 756 \\ 82 \\ +1429 \\ \hline \end{array}
\qquad
\begin{array}{r} 507 \\ 4296 \\ +39204 \\ \hline \end{array}
\qquad
\begin{array}{r} 30729 \\ 1075 \\ +298264 \\ \hline \end{array}
$$

INVERSE OPERATIONS

Many things we do can be "undone." If you take 2 steps backward, you can return to your original position by taking 2 steps forward. If you add 6 to a number, you can obtain the original number by subtracting 6 from the sum.

Any process or operation that "undoes" another process or operation is called an *inverse operation*.

Of course, there are some activities that cannot be undone. Talking cannot be undone by being silent.

Exercises 2–12:
For each activity given below, tell how to undo it.

1. Close your eyes
2. Stand up
3. Go to school
4. Close your book
5. Take 5 steps forward
6. Untie your shoe
7. Add seven
8. Subtract thirteen

SUBTRACTION

Three ways of thinking about the meaning of subtraction are explained below. The first two ways are helpful for interpreting a physical situation in terms of mathematics. However, their disadvantages will be pointed out. The last way defines subtraction for any mathematical situation.

1. Removing a subset:

John had 7 pennies and spent 4 of them. How many pennies did he have left?

We might illustrate the problem with Venn diagrams. Let $H = \{\text{pennies he had}\}$ and let $S = \{\text{pennies he spent}\}$. Each circular region in the following drawing represents a distinct penny.

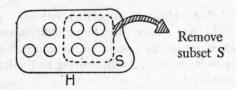

Remove subset S

Definition 2–5:

The difference between $n(H)$ and $n(S)$, denoted by $n(H) - n(S)$, is the number of members in H but not in S.

When subset S is removed from set H, only 3 pennies remain.

$$n(H) - n(S) = 3$$
$$7 - 4 = 3$$

This idea is suitable only for whole numbers and the particular type of problem illustrated.

2. Comparing sets:

Bob has 5 stamps and Jane has 9 stamps. How many more stamps does Jane have than Bob?

Let each □ in the following diagram represent a distinct stamp. Let $B = \{\text{Bob's stamps}\}$ and let $J = \{\text{Jane's stamps}\}$.

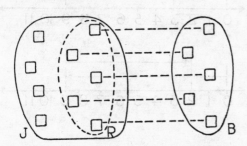

Select a subset R of J by establishing a one-to-one correspondence between set B and subset R. Since $n(R) = n(B)$, we can treat sets J and R as in the previous method.

$$n(J) = 9 \qquad n(B) = n(R) = 5$$
$$n(J) - n(B) = 4$$
$$9 - 5 = 4$$

We now have a method of treating two types of subtraction problems, but as yet subtraction is not defined for all numbers. That is, neither of the above methods is practical for fractional numbers, only for whole numbers.

3. Inverse of Addition:

By using either of the previous methods, we see that addition and subtraction are related—addition and subtraction undo each other. Addition and subtraction are inverse operations.

$$7 - 4 = 3 \text{ and } 3 + 4 = 7$$
$$9 - 5 = 4 \text{ and } 4 + 5 = 9$$

Definition 2–6:

For any numbers a, b, and c, if $c + b = a$, then c is the **difference** between a and b, denoted by $a - b$.

Note that $a - b$ names a number such that $(a - b) + b = a$. That is, we begin with the number a, subtract b, then add b, and the result is the number a with which we started. This shows the *do-undo* relationship between addition and subtraction.

Also note that $(a + b)$ names a number such that $(a + b) - b = a$. In this case subtraction undoes addition.

Exercises 2–13:

Write the simplest numeral for each of the following.

1. $(15 - 7) + 7$ 5. $57 + (756 - 57)$
2. $621 + (754 - 621)$ 6. $(39 - 17) + 17$
3. $(69 + 83) - 83$ 7. $(312 + 179) - 179$
4. $(26 - 15) + 15$ 8. $(r + s) - s$

Find the simplest numeral for each difference. Think of the corresponding addition if necessary.

9. $15 - 7$ 13. $18 - 9$
10. $11 - 8$ 14. $13 - 6$
11. $14 - 6$ 15. $12 - 5$
12. $9 - 3$ 16. $17 - 8$

FINDING UNNAMED ADDENDS

Think about solving the following problem.

Randy bought 12 pieces of candy. He ate some of them and has 5 pieces left. How many pieces of candy did he eat?

We might think: If we add the number of pieces of candy he has left (5) to the number of pieces of candy he ate ($\square$), the sum should be the number of pieces of candy he bought (12).

$$5 + \square = 12$$

Now how can we determine the simplest numeral to replace $\square$ so that $5 + \square = 12$ becomes a true sentence?

Using the inverse idea between addition and subtraction, we have

$$5 + \square = 12 \text{ so } 12 - \square = 5.$$

We see that this approach is not too helpful. We might start over again by using the commutative property of addition.

$$5 + \square = 12 \text{ so } \square + 5 = 12$$

Now use the inverse idea.

$$\square + 5 = 12 \text{ so } 12 - 5 = \square$$
$$7 = \square$$

Randy ate 7 pieces of candy.

Note that we could have used any other symbol to represent the number of pieces of candy he ate. That is, we could have used $\triangle$, $\bigcirc$, a, b, c, or any other symbol to hold a place for the numeral. Then, instead of $5 + \square = 12$, we could have written $5 + \triangle = 12$, $5 + \bigcirc = 12$, $5 + a = 12$, $5 + b = 12$, $5 + c = 12$, and so on.

Exercises 2–14:

Find the unnamed addend in each of the following.

1. $9 + \square = 15$ 4. $8 + \square = 16$
2. $7 + x = 11$ 5. $5 + k = 14$
3. $6 + n = 14$ 6. $3 + y = 12$

Write a number sentence for each problem. Solve the number sentence and write an answer for the problem.

7. A boy had 9 scout awards. He earned some more awards, and now he has 12 awards. How many more awards did he earn?

8. Jane made 15 cupcakes. Her brothers ate some of them and there are 7 left. How many of the cupcakes did her brothers eat?

9. Diane invited 12 children to her birthday party. If only 8 of the children came, how many were invited but did not attend?

10. Randy picked 7 apples from one tree and some from another tree. He picked 13 apples in all. How many did he pick from the second tree?

SUBTRACTION ON A NUMBER LINE

Recall that addition is associated with moving to the right on a number line. Since addition and subtraction are inverse operations, we expect subtraction to be associated with moving to the left on a number line.

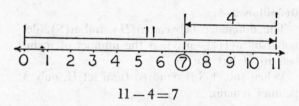

$$11 - 4 = 7$$

Exercises 2–15:

Write the subtraction sentence shown by each number-line drawing below.

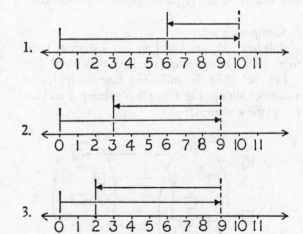

PROPERTIES OF SUBTRACTION

Is subtraction commutative? That is, can we change the order of the numbers without changing the difference?

$$7 - 3 = 4 \text{ but } 3 - 7$$

does not name a whole number, let alone being equal to 4. Hence, subtraction is *not* commutative.

Is subtraction associative? That is, can we change the grouping of the numbers without changing the difference?

$$(12 - 6) - 2 = 6 - 2 = 4$$
$$12 - (6 - 2) = 12 - 4 = 8$$

Since $4 \neq 8$, we see that subtraction is *not* associative.

ZERO IN SUBTRACTION

If the empty set (set of no members) is removed from $A = \{a,b,c,d\}$, the result is set A.

$$n(A) - n(\emptyset) = n(A)$$
$$4 - 0 \quad = 4$$

Since this is true for all sets, it is also true for all whole numbers.

If set A is removed from itself, the result is the empty set.

$$n(A) - n(A) = n(\emptyset)$$
$$4 - 4 \quad = 0$$

Since this is true for all sets, it is also true for all whole numbers.

We can summarize these two special properties of zero as follows:

For any whole number a,

$$a - 0 = a, \text{ and}$$
$$a - a = 0.$$

SUBTRACTION ALGORISM

As with addition, we should like to devise a scheme of writing the numerals for greater numbers so that subtraction can be done quickly and

easily. For example, we may want to find the difference between 758 and 326.

Follow the same procedure for still greater numbers.

```
  6975        6000 + 900 + 70 + 5
 −3864       −(3000 + 800 + 60 + 4)
 ─────       ──────────────────────
  3111        3000 + 100 + 10 + 1 = 3111
```

Exercises 2–16:

Find each difference.

1. 756	9384	67859
−531	−4150	−21536
2. 526	7925	82756
−413	−4912	−62412
3. 837	5987	49758
−216	−2852	−15047

RENAMING NUMBERS IN SUBTRACTION

Think about subtracting 592 from 857.

```
  857        800 + 50 + 7
 −592       −(500 + 90 + 2)
 ─────      ───────────────
              ? + 5
```

We notice that 90 is greater than 50, and 50–90 does not name a whole number. But we can name any number in many different ways. Let us rename 857 so that we can subtract the tens.

```
  857        700 + 150 + 7
 −592       −(500 +  90 + 2)
 ─────      ────────────────
             200 +  60 + 5 = 265
```

Another example might be the following, where we must rename 3548 so that we can subtract the ones.

$$3548 - 2419$$

$$3000 + 500 + 30 + 18$$
$$- (2000 + 400 + 10 + 9)$$
$$\overline{1000 + 100 + 20 + 9 = 1129}$$

In still other cases we find it impossible to subtract ones or tens or hundreds, and so on, or any combination of these. We merely rename the number subtracted from until subtraction becomes possible in every place-value position.

$$3426 - 1358$$

$$3000 + 400 + 10 + 16$$
$$- (1000 + 300 + 50 + 8)$$
$$\overline{? + 8}$$

Rename 3426 in another way so that subtraction of the tens is possible.

$$3426 - 1358$$

$$3000 + 300 + 110 + 16$$
$$- (1000 + 300 + 50 + 8)$$
$$\overline{2000 + 0 + 60 + 8 = 2068}$$

This procedure may be abbreviated as shown in the following examples.

$$315 - 172$$

$$200 + 110 + 5$$
$$- (100 + 70 + 2)$$
$$\overline{100 - 40 - 3}$$

$$\overset{2\ 11}{\cancel{3}\cancel{1}5} - 172$$
$$\overline{143}$$

$$752 - 328$$

$$700 + 40 + 12$$
$$- (300 + 20 + 8)$$
$$\overline{400 + 20 + 4}$$

$$7\overset{4\ 12}{\cancel{5}\cancel{2}} - 328$$
$$\overline{424}$$

Exercises 2–17:

Find each difference.

1.	315	3427	56349
	−163	−2109	−21467

2.	408	5382	47009
	−226	−3475	−20858

3.	725	6243	70000
	−537	−3856	−43197

CHECKING SUBTRACTION

Since addition and subtraction are inverse operations, we can use addition to check subtraction.

Subtraction	*Check*
715	432
−432	+283
283	715

Exercises 2–18:

Check the subtraction in Exercises 2–17.

CHAPTER THREE

MULTIPLICATION AND DIVISION OF WHOLE NUMBERS

USING SETS IN MULTIPLICATION

We have described addition of whole numbers in terms of joining disjoint sets. It is also possible to describe multiplication in this way.

A sandwich menu lists three kinds of meat—beef, ham, pork. You can have either white bread or rye bread. What are all the possible kinds of sandwiches if you choose one kind of meat and one kind of bread? The answer might be shown as follows.

(beef, white)   (ham, white)   (pork, white)
(beef, rye)   (ham, rye)   (pork, rye)

We notice that there are 3 choices of meat, 2 choices of bread, and 6 possible kinds of sandwiches. Somehow we have performed an operation on 2 and 3 to obtain 6.

Another example might involve finding the number of street intersections formed by the following situation.

{1st Ave., 2nd Ave., 3rd Ave., 4th Ave.}
{A St., B St., C St.}

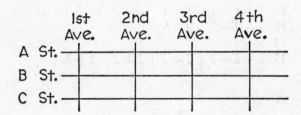

Notice that there are 4 avenues, 3 streets, and 12 intersections. Somehow we have performed an operation on 3 and 4 to obtain 12.

In a game you are to pick one letter from set A below and then pick one number from set B.

$A = \{a, b, c, d, e\}$
$B = \{1, 2, 3\}$

To show all the possible pairs we could construct the following array.

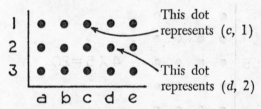

We are to choose first from set A, so let us agree to list the members of this set horizontally when making the array. We are to choose second from set B, so let us agree to list the members of this set vertically.

In other words, the number of columns in the array is the number of the first set and the number of rows is the number of the second set. In this case, the number of columns is n(A) or 5 and the number of rows is n(B) or 3.

Note that the 15 dots indicate that there are 15 possible combinations for picking one letter and one number.

Exercises 3–1:

Tell how many dots there are in the array for picking one member from the first set below and one member from the second set.

1. {Ed, Bill, Al} and {Jo, Mary, Susan}
2. {?, *} and {a, b, c, d}
3. {cake, cookies} and {coffee, tea, milk}
4. {a, b, c, d, e, f} and {7, 8, 9, 10}

DEFINITION OF MULTIPLICATION

Thinking of arrays for two sets enables us to define the operation of multiplication. Suppose we are given sets A and B such that $n(A) = a$ and $n(B) = b$. We could make an array for these sets so that it has a columns and b rows.

Definition 3–1:

The **product** of any two whole numbers a and b, denoted by $a \times b$, is the number of dots in the array having a columns and b rows. The numbers a and b are called **factors**.

The symbol $a \times b$ is read "a times b" or "the product of a and b." Hence, 4×5 is read "4 times 5" or "the product of 4 and 5." For a pictorial representation of 4×5 we can set up an array having 4 columns with 5 dots in each column, and then count the number of dots in the array.

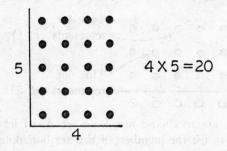

$4 \times 5 = 20$

Exercises 3–2:

Draw an array if necessary and find the simplest numeral for each product below.

1. 3×5
2. 4×6
3. 2×5
4. 7×2
5. 2×8
6. 4×4
7. 8×3
8. 5×6
9. 9×3
10. 4×7
11. 6×3
12. 3×3

MULTIPLICATION AS REPEATED ADDITION

Any array can be thought of as the union of equivalent sets.

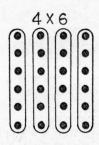

5 X 3   4 X 6

5 sets, 3 members   4 sets, 6 members
in each set   in each set

$3 + 3 + 3 + 3 + 3$ $6 + 6 + 6 + 6$
5 addends   4 addends

These drawings illustrate another way to think about multiplication of whole numbers.

$$5 \times 3 = 3 + 3 + 3 + 3 + 3 = 15$$
$$4 \times 6 = 6 + 6 + 6 + 6 = 24$$

This is sometimes referred to as the repeated addition description of multiplication.

Exercises 3–3:

Write the meaning of each product as repeated addition and give the simplest numeral for the product.

1. 6×2
2. 6×1
3. 1×6
4. 4×5
5. 5×4
6. 3×7
7. 6×7
8. 4×9
9. 5×7

Write each of the following as a product of two factors. Then give the simplest numeral for each product.

10. $8 + 8 + 8 + 8$
11. $5 + 5 + 5 + 5 + 5 + 5$
12. $1 + 1 + 1 + 1 + 1$
13. $9 + 9 + 9 + 9 + 9$
14. $9 + 9$
15. $2 + 2 + 2 + 2 + 2 + 2 + 2 + 2 + 2$

MULTIPLICATION IS COMMUTATIVE

An array having 4 columns and 3 rows can be changed into an array having 3 columns and 4 rows as shown below.

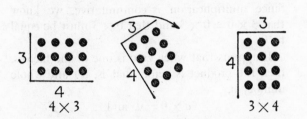

$$4 \times 3 \qquad\qquad 3 \times 4$$

Since $4 \times 3 = 12$ and $3 \times 4 = 12$, we notice that the order of the factors can be changed but the product remains the same.

That is, for all whole numbers a and b,

$$a \times b = b \times a.$$

We call this idea the **commutative property of multiplication.** Or we say that *multiplication is commutative.*

This property of multiplication can also be shown by repeated addition.

$$5 \times 4 = 4 + 4 + 4 + 4 + 4 = 20$$
$$4 \times 5 = 5 + 5 + 5 + 5 = 20$$

Therefore, $5 \times 4 = 4 \times 5$.

Exercises 3–4:

Complete each of the following sentences by using the commutative property of multiplication. Do not find any of the products.

1. $5 \times 9 = 9 \times$ ___
2. ___ $\times 8 = 8 \times 7$
3. $31 \times 7 =$ ___ $\times 31$
4. $12 \times$ ___ $= 6 \times 12$
5. $9 \times 17 =$ ___ $\times$ ___
6. $23 \times$ ___ $= 5 \times$ ___
7. ___ $\times$ ___ $= 18 \times 9$
8. ___ $\times 13 =$ ___ $\times 27$
9. $357 \times 6 =$ ___ $\times$ ___
10. ___ $\times$ ___ $= 127 \times 43$

IDENTITY NUMBER OF MULTIPLICATION

Recall that zero is the identity number of addition because for any whole number a, $a + 0 = a = 0 + a$.

We would expect the identity number of multiplication to be some number such that for any whole number a, $a \times$ ___ $= a =$ ___ $\times a$.

Each array below has but *one* column, and hence the number of dots in the array is the same as the number of dots in the single column.

$$1 \times 4 = 4 \qquad 1 \times 5 = 5 \qquad 1 \times 3 = 3$$

Or we can interpret 1×4 as using 4 as an addend only once.

$$1 \times 4 = 4$$

Since multiplication is commutative, we know that $1 \times 4 = 4 \times 1$, and we conclude that $1 \times 4 = 4 = 4 \times 1$.

Multiplying any given whole number by one, or multiplying one by any given whole number, leaves the given number unchanged. Since one is the only number with this special property, we call the number one the **identity number of multiplication.**

For any whole number a, $a \times 1 = a = 1 \times a$.

MULTIPLICATION IS ASSOCIATIVE

Recall that the pattern of the associative property of addition is
$$(a + b) + c = a + (b + c)$$
for all whole numbers a, b, and c.

Is there such a property for multiplication? Let us look at a few examples.

$$(2 \times 3) \times 4 = 6 \times 4 = 24$$
$$2 \times (3 \times 4) = 2 \times 12 = 24$$
Therefore, $(2 \times 3) \times 4 = 2 \times (3 \times 4)$.

$$(3 \times 5) \times 2 = 15 \times 2 = 30$$
$$3 \times (5 \times 2) = 3 \times 10 = 30$$
Therefore, $(3 \times 5) \times 2 = 3 \times (5 \times 2)$.

When finding the product of three numbers, we can group the first two factors or the last two factors and always get the same product.

This idea is called the **associative property of multiplication.** Or we say that *multiplication is associative.*

For all whole numbers *a*, *b*, and *c*,
$$(a \times b) \times c = a \times (b \times c).$$
We can multiply these first

or multiply these first.

Notice that when we use the associative property of multiplication, the order of the factors is *not* changed as it is when we use the commutative property of multiplication.

Exercises 3–5:

Complete each of the following sentences by using the associative property of multiplication.

1. $(3 \times 7) \times 5 = 3 \times (\underline{} \times \underline{})$
2. $4 \times (2 \times 3) = (\underline{} \times \underline{}) \times 3$
3. $(6 \times 3) \times 2 = \underline{} \times (\underline{} \times \underline{})$
4. $\underline{} \times (\underline{} \times \underline{}) = (17 \times 8) \times 5$
5. $(\underline{} \times \underline{}) \times \underline{} = 13 \times (9 \times 3)$

Each of the following is true because of the commutative property of multiplication, the associative property of multiplication, or both of these properties. Write the letter C, A, or both C and A to tell which property or properties are used.

6. $(9 \times 8) \times 3 = 9 \times (8 \times 3)$
7. $(9 \times 8) \times 3 = 3 \times (9 \times 8)$
8. $6 \times (7 \times 12) = 6 \times (12 \times 7)$
9. $6 \times (7 \times 12) = (6 \times 12) \times 7$
10. $(13 \times 5) \times 14 = 14 \times (5 \times 13)$
11. $(32 \times 9) \times 8 = 9 \times (32 \times 8)$
12. $r \times (s \times t) = (r \times s) \times t$

ZERO IN MULTIPLICATION

What number is named by 3×0? By thinking of repeated addition,
$$3 \times 0 = 0 + 0 + 0 = 0.$$
Since multiplication is commutative, we know that $3 \times 0 = 0 \times 3$ and that 0×3 must be equal to 0.

It appears that when zero is one of the factors, then the product is zero. That is, for any whole number *a*,
$$a \times 0 = 0, \text{ and}$$
$$0 \times a = 0.$$
What can we say about the factors if the product is zero? That is, what do we know about the factors *a* and *b* if $a \times b = 0$? The only way we can get a product of zero is to use zero as one of the factors. That is:

If $a \times b = 0$, then $a = 0$, $b = 0$, or both factors are zero.

THE DISTRIBUTIVE PROPERTY

Four boys and three girls are planning a party. Each child is to bring 2 gifts. How many gifts did they bring in all?

Two ways of thinking about solving this problem are given below.

1. There are $4 + 3$ or 7 children. Each child will bring 2 gifts. Then, all together they will bring
$$(4 + 3) \times 2 \text{ or } 7 \times 2 \text{ or } 14 \text{ gifts.}$$

2. Each of the 4 boys will bring 2 gifts. Then the boys will bring 4×2 gifts. Each of the 3 girls will bring 2 gifts. Then the girls will bring 3×2 gifts. All together the children will bring
$$(4 \times 2) + (3 \times 2) \text{ or } 8 + 6 \text{ or } 14 \text{ gifts.}$$

From these ways of thinking about the problem we see that
$$(4 + 3) \times 2 = (4 \times 2) + (3 \times 2).$$
Since multiplication is commutative, we know that we can change the order of the factors. Hence,
$$(4 + 3) \times 2 = 2 \times (4 + 3),$$
$$4 \times 2 = 2 \times 4, \text{ and}$$
$$3 \times 2 = 2 \times 3.$$

Then the sentence
$$(4 + 3) \times 2 = (4 \times 2) + (3 \times 2)$$
can be written
$$2 \times (4 + 3) = (2 \times 4) + (2 \times 3).$$

Let us investigate such a pattern with different numbers.

$2 \times (5+3) = 2 \times 8 = 16$
$(2 \times 5) + (2 \times 3) = 10 + 6 = 16$
Hence $2 \times (5+3) = (2 \times 5) + (2 \times 3)$.

$(4+6) \times 3 = 10 \times 3 = 30$
$(4 \times 3) + (6 \times 3) = 12 + 18 = 30$
Hence, $(4+6) \times 3 = (4 \times 3) + (6 \times 3)$.

This pattern is also visible in an array.

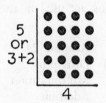

$4 \times (3+2)$ $(4 \times 3) + (4 \times 2)$
$4 \times (3+2) = (4 \times 3) + (4 \times 2)$

By drawing a horizontal line in the array we can separate it into two arrays, one having 2 rows and 4 columns and the other having 3 rows and 4 columns. We have not discarded any of the dots, so the number of dots remains the same.

This property is called the **distributive property of multiplication over addition.**

For all whole numbers a, b, and c,
$$a \times (b+c) = (a \times b) + (a \times c)$$
and
$$(b+c) \times a = (b \times a) + (c \times a).$$

It is important that we realize that the distributive property involves both addition and multiplication. Furthermore, it is important that we are able to "undistribute" as follows.

$(5 \times 3) + (5 \times 6) = 5 \times (3+6)$
$(4 \times 6) + (7 \times 6) = (4+7) \times 6$

Exercises 3–6:

Use the distributive property to complete each of the following sentences.

1. $7 \times (2+5) = (7 \times \underline{\ \ }) + (7 \times \underline{\ \ })$
2. $4 \times (3+6) = (\underline{\ \ } \times \underline{\ \ }) + (\underline{\ \ } \times \underline{\ \ })$
3. $(4+5) \times 3 = (\underline{\ \ } \times 3) + (\underline{\ \ } \times 3)$
4. $(8+9) \times 6 = (\underline{\ \ } \times \underline{\ \ }) + (\underline{\ \ } \times \underline{\ \ })$
5. $(5 \times 2) + (7 \times 2) = (\underline{\ \ } + \underline{\ \ }) \times 2$
6. $(6 \times 3) + (8 \times 3) = (\underline{\ \ } + \underline{\ \ }) \times \underline{\ \ }$

BASIC MULTIPLICATION FACTS

We can use addition or an array to determine and memorize the basic multiplication facts as shown in the following multiplication table. The first factor is named in the left column and the second factor is named in the top row.

×	0	1	2	3	4	5	6	7	8	9
0	0									
1	0	1								
2	0	2	4							
3	0	3	6	9						
4	0	4	8	12	16					
5	0	5	10	15	20	25				
6	0	6	12	18	24	30	36			
7	0	7	14	21	28	35	42	49		
8	0	8	16	24	32	40	48	56	64	
9	0	9	18	27	36	45	54	63	72	81

Since multiplication is commutative, we need not compute products for the shaded portion of the table. If we need to compute 3×7, we merely commute, $3 \times 7 = 7 \times 3$, and use the table to find that $7 \times 3 = 21$.

Even some of the basic facts can be found in other ways. For example:

$7 \times 8 = 7 \times (3+5)$	Rename 8 as $3+5$
$= (7 \times 3) + (7 \times 5)$	Dist. prop.
$= 21 + 35$	Multiplication
$= 56$	Addition

$9 \times 5 = (5+4) \times 5$	Rename 9 as $5+4$
$= (5 \times 5) + (4 \times 5)$	Dist. prop.
$= 25 + 20$	Multiplication
$= 45$	Addition

FACTORS OF 10, 100, OR 1000

When finding the product of two factors we might draw the array and count the dots. Or we might restate the multiplication as repeated addition and find the sum. Either of these methods become time-consuming and tedious when the factors are greater numbers, such as 615×352. Hence, we should like to discover some "streamlined" way of finding a product.

Let us begin by investigating products where one of the factors is 1, 10, 100, or 1000.

$$5 \times 1 = 1 + 1 + 1 + 1 + 1 = 5$$
$$5 \times 10 = 10 + 10 + 10 + 10 + 10 = 50$$
$$5 \times 100 = 100 + 100 + 100 + 100 + 100 = 500$$
$$5 \times 1000 = 1000 + 1000 + 1000 + 1000 + 1000 = 5000$$

$$7 \times 1 = 1 + 1 + 1 + 1 + 1 + 1 + 1 = 7$$
$$7 \times 10 = 10 + 10 + 10 + 10 + 10 + 10 + 10 = 70$$
$$7 \times 100 = 100 + 100 + 100 + 100 + 100 + 100 + 100 = 700$$
$$7 \times 1000 = 1000 + 1000 + 1000 + 1000 + 1000 + 1000 + 1000 = 7000$$

From these examples it is evident that to multiply by 10, annex one zero; to multiply by 100, annex two zeros; to multiply by 1000, annex three zeros.

Exercises 3–7:

Write the simplest numeral for each product.

1. 6×10	6. 4×10
2. 9×100	7. 4×100
3. 3×10	8. 4×1000
4. 8×100	9. 100×4
5. 2×100	10. 1000×4

Now let us investigate the case where one or both factors are multiples of a power of ten, such as 30, 400, 70, or 900.

$$
\begin{aligned}
8 \times 20 &= 8 \times (2 \times 10) & \text{Rename 20} \\
&= (8 \times 2) \times 10 & \text{Assoc. prop.} \\
&= 16 \times 10 & \text{Multiplication} \\
&= 160 & \text{Mult. by 10}
\end{aligned}
$$

$$
\begin{aligned}
50 \times 70 &= (5 \times 10) \times (7 \times 10) & \text{Rename factors} \\
&= [(5 \times 10) \times 7] \times 10 & \text{Assoc. prop.} \\
&= [7 \times (5 \times 10)] \times 10 & \text{Comm. prop.} \\
&= [(7 \times 5) \times 10] \times 10 & \text{Assoc. prop.} \\
&= (7 \times 5) \times (10 \times 10) & \text{Assoc. prop.} \\
&= 35 \times 100 & \text{Multiplication} \\
&= 3500 & \text{Mult. by 100}
\end{aligned}
$$

Exercises 3–8:

Write the simplest numeral for each product.

1. 7×30	3. 50×90	5. 40×600
2. 8×400	4. 70×20	6. 700×200

TECHNIQUES OF MULTIPLICATION

To find products of greater numbers, we simply use the properties in such a way that calculation is easy. Notice how the renaming of numbers and the properties of the operations are used in the following examples.

$$
\begin{aligned}
8 \times 24 &= 8 \times (20 + 4) & \text{Rename 24} \\
&= (8 \times 20) + (8 \times 4) & \text{Dist. prop.} \\
&= 160 + 32 & \text{Multiplication} \\
&= 192 & \text{Addition}
\end{aligned}
$$

$$
\begin{aligned}
37 \times 6 &= (30 + 7) \times 6 & \text{Rename 37} \\
&= (30 \times 6) + (7 \times 6) & \text{Dist. prop.} \\
&= 180 + 42 & \text{Multiplication} \\
&= 222 & \text{Addition}
\end{aligned}
$$

To make the addition easier, we can write the above example as follows.

$$
\begin{array}{r}
37 \\
\times 6 \\
\hline
42 \quad = (7 \times 6) \\
180 \quad = (30 \times 6) \\
\hline
222 \quad = (30 \times 6) + (7 \times 6)
\end{array}
$$

The distributive property of multiplication over addition is used as the factors become greater.

$$
\begin{aligned}
372 \times 4 &= (300 + 70 + 2) \times 4 & \text{Rename 372} \\
& & \text{Dist. prop.} \\
&= (300 \times 4) + (70 \times 4) + (2 \times 4) \\
& & \text{Multiplication} \\
&= 1200 + 280 + 8 \\
& & \text{Addition} \\
&= 1488
\end{aligned}
$$

This same product can be computed as follows.

$$
\begin{array}{r}
372 \\
\times 4 \\
\hline
8 \quad = (2 \times 4) \\
280 \quad = (70 \times 4) \\
1200 \quad = (300 \times 4) \\
\hline
1488
\end{array}
$$

Exercises 3–9:

Find the simplest numeral for each product.

1. 56×3 4. 354×6

2. 78×5 5. 37×8

3. 124×6 6. 426×9

7.   47   9. 173   11. 457
    $\times 4$ $\times 6$ $\times 3$

8.   29   10. 321   12. 634
    $\times 5$ $\times 8$ $\times 9$

Now let us see what happens when both factors are named by two-digit numerals.

$$24 \times 63 = 24 \times (60 + 3) \qquad \text{Rename 63}$$
$$\qquad\qquad\qquad\qquad\qquad \text{Dist. prop.}$$
$$= (24 \times 60) + (24 \times 3)$$
$$\qquad\qquad\qquad\qquad\qquad \text{Rename 60}$$
$$= [24 \times (6 \times 10)] + (24 \times 3)$$
$$\qquad\qquad\qquad\qquad\qquad \text{Assoc. prop.}$$
$$= [(24 \times 6) \times 10] + (24 \times 3)$$
$$\qquad\qquad\qquad\qquad\qquad \text{Multiplication}$$
$$= (144 \times 10) + (24 \times 3)$$
$$\qquad\qquad\qquad\qquad\qquad \text{Mult. by 10}$$
$$= 1440 + 72$$
$$\qquad\qquad\qquad\qquad\qquad \text{Addition}$$
$$= 1512$$

Another way to think about 24×63 is shown below.

$$\qquad\qquad\qquad\qquad\qquad \text{Rename factors}$$
$$24 \times 63 = (20 + 4) \times (60 + 3)$$
$$\qquad\qquad\qquad\qquad\qquad \text{Dist. prop.}$$
$$= [(20 + 4) \times 60] + [(20 + 4) \times 3]$$
$$\qquad\qquad\qquad\qquad\qquad \text{Dist. prop.}$$
$$= [(20 \times 60) + (4 \times 60)] + (20 \times 3) + (4 \times 3)$$
$$\qquad\qquad\qquad\qquad\qquad \text{Assoc. prop.}$$
$$= (20 \times 60) + (4 \times 60) + (20 \times 3) + 4 \times 3)$$
$$\qquad\qquad\qquad\qquad\qquad \text{Multiplication}$$
$$= 1200 + 240 + 60 + 12$$
$$\qquad\qquad\qquad\qquad\qquad \text{Addition}$$
$$= 1512$$

This, too, can be shown in vertical arrangement.

$$
\begin{array}{rl}
24 & \\
\times 63 & \\
\hline
12 & = (4 \times 3) \\
60 & = (20 \times 3) \\
240 & = (4 \times 60) \\
1200 & = (20 \times 60) \\
\hline
1512 &
\end{array}
$$

Exercises 3–10:

Find the simplest numeral for each product.

1.   38   4.   24   7.   83
  $\times 13$ $\times 54$ $\times 49$

2.   27   5.   78   8.   57
  $\times 24$ $\times 52$ $\times 68$

3.   47   6.   17   9.   25
  $\times 35$ $\times 58$ $\times 93$

THE MULTIPLICATION ALGORISM

The previous exercises enable us to find the product of still greater factors. You should have noticed by now that the distributive property of multiplication over addition is a powerful tool in making multiplication easy.

$$
\begin{array}{rl}
523 & \\
\times 7 & \\
\hline
21 & = (3 \times 7) \\
140 & = (20 \times 7) \\
3500 & = (500 \times 7) \\
\hline
3661 &
\end{array}
$$

$$
\begin{array}{rl}
346 & \\
\times 24 & \\
\hline
24 & = (6 \times 4) \\
160 & = (40 \times 4) \\
1200 & = (300 \times 4) \\
120 & = (6 \times 20) \\
800 & = (40 \times 20) \\
6000 & = (300 \times 20) \\
\hline
8304 &
\end{array}
\qquad
\begin{array}{r}
346 \\
\times 24 \\
\hline
1384 \\
6920 \\
\hline
8304
\end{array}
$$

By doing 346×4 mentally, we can write 1384 in the arrangement at the right. Then write 6920 by doing 346×20 mentally.

$$
\begin{array}{rl}
243 & \\
\times\, 312 & \\
\hline
486 & = (243 \times 2) \\
2430 & = (243 \times 10) \\
72900 & = (243 \times 30) \\
\hline
75816 &
\end{array}
$$

We can erase the indicated products at the right, and omit writing the final 0's of 2430 and 72,900, and write the short form as follows.

$$
\begin{array}{r}
243 \\
\times 312 \\
\hline
486 \\
243 \\
729 \\
\hline
75816
\end{array}
$$

Exercises 3–11:
Find each product.

1.	624 ×3	4.	432 ×312	7.	3426 ×4	
2.	532 ×24	5.	675 ×123	8.	4521 ×32	
3.	324 ×31	6.	843 ×324	9.	3421 ×322	

ESTIMATING A PRODUCT

Often we are interested in estimates rather than the exact answers. Knowing how to multiply by multiples of powers of ten helps us find an estimate of a product very quickly and easily. Suppose you are to find the value of n in $n = 28 \times 53$.

$$20 < 28 \text{ and } 50 < 53,$$
so $20 \times 50 < n$ or $1000 < n.$
$$30 > 28 \text{ and } 60 > 53,$$
so $30 \times 60 > n$ or $1800 > n.$

Hence, we know that $1000 < n < 1800$, which is read: 1000 is less than n *and* n is less than 1800. Another way of saying this is "n is between 1000 and 1800."

Another example might be to find the value of x in $x = 72 \times 587$.

$$70 < 72 \text{ and } 500 < 587,$$
so 70×500 or $35000 < x.$
$$80 > 72 \text{ and } 600 > 587,$$
so 80×600 or $48000 > x.$

Hence, $35000 < x < 48000$, or the value of x is between 35000 and 48000.

Exercises 3–12:
Find a rough estimate for n in each sentence. Then find the exact answer.

1. $n = 27 \times 65$ 5. $n = 47 \times 367$
2. $n = 82 \times 75$ 6. $n = 77 \times 492$
3. $n = 39 \times 58$ 7. $n = 826 \times 52$
4. $n = 43 \times 94$ 8. $n = 572 \times 67$

DIVISION

Division is related to multiplication in much the same way that subtraction is related to addition. When two numbers are added, the addition can be undone by subtraction. Similarly, when two numbers are multiplied, the multiplication can be undone by division. Hence, multiplication and division are inverse operations.

Addition	*Subtraction*
$5 + 7 = 12$	$12 - 7 = 5$
$29 + 5 = 34$	$34 - 5 = 29$

Multiplication	*Division*
$7 \times 6 = 42$	$42 \div 6 = 7$
$24 \times 3 = 72$	$72 \div 3 = 24$

Knowing that multiplication and division are inverse operations, we can interpret $8 \div 2$ as that factor which, when multiplied by 2, yields a product of 8. That is,

$$(8 \div 2) \times 2 = 8.$$

If we should think of an array as we did for multiplication, $8 \div 2$ would be the number of columns in an array of 8 dots having 2 dots in each column.

$$8 \div 2 \text{ or } 4$$

Or we can think of $8 \div 2$ as the number of disjoint subsets formed when a set of 8 objects is separated into disjoint subsets having 2 objects each.

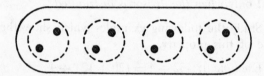

$$8 \div 2 = 4$$

Set of 8 objects separated into 4 disjoint subsets having 2 objects each.

Exercises 3–13:

Think of an array or separating a set into disjoint equivalent subsets to tell the number named by each of the following.

1. $12 \div 3$	5. $20 \div 4$	9. $24 \div 8$
2. $6 \div 2$	6. $18 \div 2$	10. $24 \div 12$
3. $15 \div 5$	7. $14 \div 7$	11. $24 \div 3$
4. $16 \div 4$	8. $24 \div 6$	12. $24 \div 4$

ZERO IN DIVISION

First let us investigate a division such as $7 \div 0 = n$. Since multiplication and division are inverse operations, the above division can be restated as a multiplication.

$$7 \div 0 = n \text{ so } n \times 0 = 7$$

But we already know that when one of the factors is zero, the product is zero. Hence, there is *no* number n such that $n \times 0 = 7$. That is, $7 \div 0$ does not name a number.

Now let us investigate the special case $0 \div 0 = n$. Restate this as a multiplication.

$$0 \div 0 = n \text{ so } n \times 0 = 0$$

In this case, any number we choose for n yields a product of 0. That is, $5 \times 0 = 0$, $721 \times 0 = 0$,

$9075 \times 0 = 0$, and so on. If we accept $0 \div 0$ as a name for a number, then we are forced to accept that it names *every* number. This is certainly not very helpful.

The fact that in the first case *no* number is named and in the second case *every* number is named is a source of difficulty in division. Let us rule out both of these cases by agreeing to the following.

Division by zero is meaningless. This means that we shall not define division by zero.

Now let us investigate a case such as $0 \div 8 = n$. Restate this as a multiplication.

$$0 \div 8 = n \text{ so } n \times 8 = 0$$

We already know that if the product is zero, at least one of the factors must be zero. Since $8 \neq 0$, then n must be equal to zero. Hence, $0 \times 8 = 0$ and $0 \div 8 = 0$.

This is true regardless of what number we choose for a, except $a = 0$, in the following.

$$0 \div a = 0 \text{ if } a \neq 0.$$

We can state our finding as follows. When zero is divided by any nonzero number the result is zero.

DEFINITION OF DIVISION

Now let us state a definition of division.

Definition 3–2:

If $a \times b = c$ and $b \neq 0$, then $c \div b = a$.

We read $c \div b$ as "c divided by b."

The number named by $c \div b$ is called the **quotient,** the number named by b is called the **divisor,** and the number named by c is called the **dividend.**

We can show the relationship between these numbers and the numbers in a multiplication as follows.

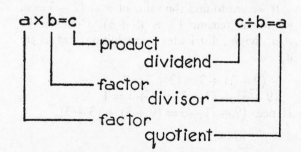

Hence, we see that in a division we are given the product and one of the factors, and we are to find the other factor.

From this definition, and knowing the basic multiplication facts, we can determine the basic definition facts.

Exercises 3–14:

Find each quotient.

1. $42 \div 7$	5. $36 \div 4$	9. $25 \div 5$
2. $35 \div 5$	6. $21 \div 3$	10. $49 \div 7$
3. $48 \div 8$	7. $45 \div 9$	11. $27 \div 3$
4. $81 \div 9$	8. $32 \div 8$	12. $28 \div 7$

PROPERTIES OF DIVISION

Is division commutative? That is, do $12 \div 4$ and $4 \div 12$ name the same number? $12 \div 4 = 3$, but $4 \div 12$ does not name a whole number, let alone three. Hence, $12 \div 4 \neq 4 \div 12$. Division is *not* commutative.

Is division associative? That is, do $(12 \div 6) \div 2$ and $12 \div (6 \div 2)$ name the same number?

$$(12 \div 6) \div 2 = 2 \div 2 = 1$$
$$12 \div (6 \div 2) = 12 \div 3 = 4$$

Hence, $(12 \div 6) \div 2 \neq 12 \div (6 \div 2)$. Division is *not* associative.

Since division is not commutative, we know that $8 \div 1 \neq 1 \div 8$. But let us see what happens when the divisor is 1.

$$8 \div 1 = n \text{ so } n \times 1 = 8$$

Since 1 is the identity number of multiplication, we see that $n = 8$ and $8 \div 1 = 8$. That is, when the divisor is 1, the dividend and the quotient are the same.

For all whole numbers a,

$$a \div 1 = a.$$

If we are to find the value of n in $12 \div 3 = n$, we might rename 12 as $(9 + 3)$. Could it be that division distributes over addition? Let us try it.

$$(9 + 3) \div 3 = 12 \div 3 = 4$$
$$(9 \div 3) + (3 \div 3) = 3 + 1 = 4$$

Hence, $(9 + 3) \div 3 = (9 \div 3) + (3 \div 3)$.

We might be tempted to try the other pattern of the distributive property. That is, rename 3 as $2 + 1$ and write $12 \div 3$ as $12 \div (2 + 1)$.

$$12 \div (2 + 1) = 12 \div 3 = 4$$
$$(12 \div 2) + (12 \div 1) = 6 + 12 = 18$$

Hence, $12 \div (2 + 1) \neq (12 \div 2) + (12 \div 1)$.

However, it is important to remember that for all whole numbers a, b, and c, where $c \neq 0$,

$$(a + b) \div c = (a \div c) + (b \div c).$$

We say that *division distributes over addition, but only when the divisor is distributed.*

Study the following examples which show the use of this property.

Example 1:
$$\begin{aligned} 32 \div 4 &= (20 + 12) \div 4 \\ &= (20 \div 4) + (12 \div 4) \\ &= \quad 5 \quad + \quad 3 \\ &= \quad 8 \end{aligned}$$

Example 2:
$$\begin{aligned} 75 \div 5 &= (40 + 35) \div 5 \\ &= (40 \div 5) + (35 \div 5) \\ &= \quad 8 \quad + \quad 7 \\ &= \quad 15 \end{aligned}$$

Example 3:
$$\begin{aligned} 75 \div 5 &= (50 + 25) \div 5 \\ &= (50 \div 5) + (25 \div 5) \\ &= \quad 10 \quad + \quad 5 \\ &= \quad 15 \end{aligned}$$

Exercises 3–15:

Rename the dividend in each of the following and use the distributive property of division over addition to find each quotient.

1. $16 \div 2$	4. $44 \div 4$	7. $52 \div 4$
2. $65 \div 5$	5. $84 \div 7$	8. $72 \div 6$
3. $39 \div 3$	6. $24 \div 4$	9. $95 \div 5$

REMAINDERS IN DIVISION

If we think of separating a set into disjoint equivalent subsets, we find that some divisions do not yield a whole number as a quotient.

12 objects, 3 disjoint sets,
4 objects in each subset
$12 \div 4 = 3$

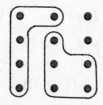

12 objects, 2 disjoint subsets
of 5 objects each, *and* 2 objects
left over

$$12 = (2 \times 5) + 2$$
$$\uparrow \qquad\qquad \uparrow$$
quotient   remainder

As long as we are operating only with whole numbers, we shall give the remainder as such. Later in this book we will extend the number system so that we can carry out division without having to use remainders.

A final comment regarding the remainder is the following.

$$20 = (4 \times 5) + 0,$$
so $20 \div 5 = 4$ with remainder 0.

Let us agree that every division of whole numbers has a remainder. That is, for whole numbers a and b, $b \neq 0$, the division $a \div b$ can be stated as

$$a = (q \times b) + r$$

where q is the quotient and r is the remainder. Furthermore, $r = 0$ or $r > 0$ and $r < b$. This means that the remainder is either zero or some whole number between 0 and b.

Exercises 3–16:

Find each quotient and remainder.

1. $17 \div 5$ 4. $31 \div 5$ 7. $55 \div 4$
2. $21 \div 6$ 5. $40 \div 9$ 8. $73 \div 8$
3. $15 \div 7$ 6. $33 \div 6$ 9. $67 \div 6$

ONE-DIGIT DIVISORS

The definition of division does not tell us how to carry out a division. We should like to discover some method of writing the numerals so that division becomes easy, especially where greater numbers are involved.

Suppose we want to find the simplest numeral for $43 \div 5$. In terms of multiplication we can estimate the result by thinking of $n \times 5 = 43$. Knowing the multiples of 5 helps us in making this estimate.

$$8 \times 5 = 40 \text{ and } 40 < 43$$
$$9 \times 5 = 45 \text{ and } 45 > 43$$

Obviously, $43 \div 5$ does not name a whole number. Perhaps we can state the result in the form $a = (q \times b) + r$. So let us rename 43 as a sum of two addends, the first of which is 40 since we already know it can be named as 8×5.

$$43 = 40 + 3$$
$$= (8 \times 5) + 3$$
$$\uparrow \qquad\qquad \uparrow$$
quotient   remainder

Another way of writing this is:

$$\begin{array}{r} 8 \\ 5\overline{)43} \\ \underline{40} = (8 \times 5) \\ 3 \end{array} \quad \text{or} \quad \begin{array}{r} 8 \\ 5\overline{)43} \\ \underline{40} \\ 3 \end{array}$$

Now let us find the simplest numeral for $256 \div 8$.

Think of $n \times 8 = 256$ to estimate the result.

$$10 \times 8 = 80 \text{ and } 80 < 256$$
$$20 \times 8 = 160 \text{ and } 160 < 256$$
$$30 \times 8 = 240 \text{ and } 240 < 256$$
$$40 \times 8 = 320 \text{ and } 320 > 256$$

Hence, the result is between 30 and 40. So let us rename 256 as the sum of two addends, the first addend being 240.

$$256 = 240 + 16$$

We notice that the second addend cannot be the remainder since $16 > 8$. Further, we notice that 16 is a multiple of 8. Then we can use the distributive property of division over addition.

$$256 \div 8 = (240 + 16) \div 8$$
$$= (240 \div 8) + (16 \div 8)$$
$$= \quad 30 \quad + \quad 2$$
$$= 32$$

Another way to write this is:

$$30 + 2 = 32$$
$$8\overline{)256} \quad 8\overline{)240 + 16}$$

Or a more concise method is to think of place value.

$$
\begin{array}{r}
3 \\
8\overline{)256} \\
240 \\
\hline
16
\end{array}
\quad = (30 \times 8)
\qquad
\begin{array}{r}
32 \\
8\overline{)256} \\
240 \\
\hline
16 \\
16 \\
\hline
0
\end{array}
\quad = (2 \times 8)
$$

Exercises 3–17:

Find each quotient.

1. $105 \div 7$ 4. $315 \div 5$ 7. $3320 \div 8$
2. $156 \div 6$ 5. $288 \div 4$ 8. $2526 \div 6$
3. $272 \div 8$ 6. $201 \div 3$ 9. $2464 \div 7$

TWO-DIGIT DIVISORS

Let us extend division to the case where the divisor is expressed by a two-digit numeral. For example, let us carry out the division $624 \div 32$.

We are not as familiar with the multiples of 32 as we are with the multiples of the numbers 1 through 10, so let us think of renaming 624 as the sum of more than two addends. Let us do this by thinking of multiples of powers of 10 as shown below.

$$10 \times 32 = 320 \text{ and } 320 < 624$$
$$20 \times 32 = 640 \text{ and } 640 > 624$$

Now we can rename 624 as follows.

$$624 = 320 + 304$$
$$= (10 \times 32) + 304$$

Certainly 304 cannot be the remainder since $304 > 32$. Now rename 304 as a sum of two addends where the first addend is a multiple of 32. Since 304 is nearly 320 we can expect the value of n in $n \times 32 = 304$ to be nearly 10.

$$8 \times 32 = 256 \text{ and } 256 < 304$$
$$9 \times 32 = 288 \text{ and } 288 < 304$$
$$10 \times 32 = 320 \text{ and } 320 > 304$$

Now we can rename 624 so that the division can be completed easily.

$$624 = 320 + 288 + 16$$
$$= (10 \times 32) + (9 \times 32) + 16$$
$$= (19 \times 32) + 16$$
$$\qquad \uparrow \qquad\qquad \uparrow$$
$$\text{quotient} \qquad \text{remainder}$$

Again, to make the subtraction (necessary in the renaming) easy, let us proceed as follows.

$$
\begin{array}{cc}
Think & Write
\end{array}
$$

$$
\left.\begin{array}{r} 9 \\ 10 \end{array}\right\} 10 + 9 = 19 \qquad 19
$$

$$
\begin{array}{r}
32\overline{)624} \\
320 \\
\hline
304 \\
288 \\
\hline
16
\end{array}
\quad
\begin{array}{l}
= (10 \times 32) \\
\\
= (9 \times 32)
\end{array}
\qquad
\begin{array}{r}
32\overline{)624} \\
320 \\
\hline
304 \\
288 \\
\hline
16
\end{array}
$$

Exercises 3–18:

Carry out each division.

1. $415 \div 25$ 4. $328 \div 45$ 7. $2091 \div 17$
2. $534 \div 31$ 5. $408 \div 24$ 8. $2115 \div 17$
3. $715 \div 64$ 6. $638 \div 53$ 9. $3400 \div 17$

THE DIVISION ALGORISM

The vertical method makes the division easier to complete since it makes subtraction easy. However, we are faced with subtracting the greatest multiple of the divisor. Suppose we illustrate this with $329 \div 14$.

$$
\begin{array}{r}
1 \\
14\overline{)329} \\
140 \\
\hline
189
\end{array}
\quad = (10 \times 14)
$$

We see that $189 > (10 \times 14)$, so 10×14 is not the greatest multiple of 14 that we could have subtracted in this process. So let us start over again.

$$
\begin{array}{r}
2 \\
14\overline{)329} \\
280 \\
\hline
49
\end{array}
\quad = (20 \times 14)
$$

Now estimate the value of n in $n \times 14 = 49$ to obtain the next digit in the answer.

$$3 \times 14 = 42 \text{ and } 42 < 49$$
$$4 \times 14 = 56 \text{ and } 56 > 49$$

Hence, the next digit in the answer must be **3**.

Think		*Write*
23		23
14)329		14)329
280	$= (20 \times 14)$	280
49		49
42	$= (3 \times 14)$	42
7		7

Study the following examples and notice how this process is used.

$$
\begin{array}{r}
167 \\
21\overline{)3526} \\
2100 \quad = (100 \times 21) \\
\overline{1426} \\
1260 \quad = (60 \times 21) \\
\overline{166} \\
147 \quad = (7 \times 21) \\
\overline{19}
\end{array}
\qquad
\begin{array}{r}
81 \\
58\overline{)4708} \\
4640 \quad = (80 \times 58) \\
\overline{68} \\
58 \quad = (1 \times 58) \\
\overline{10}
\end{array}
$$

Exercises 3–19:

Carry out each division.

1. $217 \div 34$
2. $836 \div 22$
3. $759 \div 18$
4. $342 \div 32$
5. $5806 \div 47$
6. $7052 \div 35$
7. $3431 \div 25$
8. $5284 \div 25$
9. $8059 \div 25$
10. $6072 \div 253$
11. $2448 \div 24$
12. $6794 \div 79$

CHAPTER FOUR

THE SET OF INTEGERS

CLOSURE UNDER ADDITION

You may have observed that the sum of two whole numbers is always a whole number. For example, $6 + 2 = 8$, $29 + 15 = 44$, and so on. Never is the sum different from a whole number, such as $3\frac{1}{2}$ or $7\frac{3}{4}$. We express this idea by saying that the set of whole numbers is *closed* under addition, or closed with respect to addition.

Notice that this idea involves both a set of numbers and an operation. Hence, we cannot say that closure is a property of a set or of an operation, but depends on both of them.

Definition 4–1:

If a and b are any two numbers (not necessarily different) of set A and $+$ denotes the operation addition, then set A is closed under addition if $a + b$ is a number of set A.

In the above case, set A is the set of whole numbers, and by the definition we can say that the set of whole numbers is closed under addition.

Is the set of whole numbers closed under subtraction? If it is, then for all whole numbers a and b, $a - b$ should be a whole number. But we already know that $3 - 5$, $8 - 36$, and so on, do not name whole numbers. Thus, the set of whole numbers is *not closed* under subtraction.

In other words, we can state certain subtraction problems, or write certain equations, such as $5 + n = 0$ or $4 - x = 7$, which cannot be solved by using only the set of whole numbers.

THE INTEGERS

Consider the equation $5 + n = 0$. Since there is no such whole number n for which $5 + n = 0$ is true, let us invent one. Let us call this new number *the negative of five* and denote it by the symbol -5. Then we can write $5 + (-5) = 0$, since -5 was invented especially for this purpose.

Obviously, for every whole number a (except 0) we can invent a new number $-a$ so that $a + (-a) = 0$. Zero is excluded because $0 + (-0) = 0$ is equivalent to $0 + 0 = 0$; and why burden ourselves with two symbols for the number zero?

Definition 4–2:

The set of *integers* is the union of the set of whole numbers and all of these newly invented numbers.
The integers: $\{\ldots, -4, -3, -2, -1, 0, 1, 2, 3, \ldots\}$

The set of numbers $\{\ldots, -4, -3, -2, -1\}$ is called the set of *negative integers*. The set of numbers $\{1, 2, 3, \ldots\}$ is called the set of *positive integers*. Furthermore, the set of numbers $\{0, 1, 2, 3, \ldots\}$ is called the set of *nonnegative integers*, and the set of numbers $\{\ldots, -3, -2, -1, 0\}$ is called the set of *nonpositive integers*.

The symbol $-$ is used in two ways. One way is to indicate the negative of a number, such as in -7. The other way is to indicate subtraction, such as in $9 - 4$. Later we will see a definite relationship between these two uses and no confusion will arise.

Exercises 4–1:

Which numeral should replace x in each of the following so that each sentence becomes true?

1. $7 + x = 0$ 4. $x + (12) = 0$
2. $15 + x = 0$ 5. $129 + (-129) = x$
3. $9 + (-9) = x$ 6. $306 + x = 0$

THE NUMBER LINE

When using the number line to show addition of whole numbers, you may have wondered about extending the number line to the left. Certainly we can locate points to the left of the 0-point just as we do to the right of the 0-point. Let us label them with the numerals for the negative integers, as shown below.

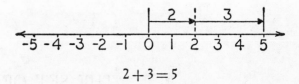

Since a number (such as 2) and its negative (such as -2) are on opposite sides of the 0-point, you may think of them as being *opposites* of each other. Futhermore, note that the distance from the 0-point to the point for any given number is the same as the distance from the 0-point to the opposite of the given number.

The integers are sometimes referred to as *directed numbers* because they not only indicate "how many" but also in "what direction."

Many life situations serve to make us aware of the need and use of such numbers. We refer to temperature readings above zero and below zero. Hence, a change of direction is implied by the words "above" and "below."

Or someone gains $100 on a business transaction and loses $50 on another business transaction. In this case, the change of direction is indicated by the words "gain" and "loss" and the dollar amounts indicate the extent of the gain or loss.

ADDITION ON A NUMBER LINE

Now let us investigate the use of moves on a number line to help us understand the addition

of integers. The sum of two positive integers is shown in the same way as the sum of two whole numbers.

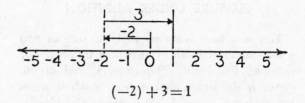

$$2 + 3 = 5$$

The following number line shows how to find the sum of -2 and 3.

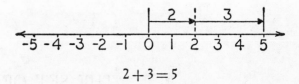

$$(-2) + 3 = 1$$

Since addition of integers is commutative, we know that $3 + (-2) = 1$, as shown on the following number line.

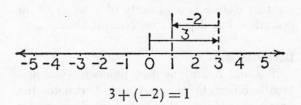

$$3 + (-2) = 1$$

Another example of this type is $(-5) + 3$, as shown below.

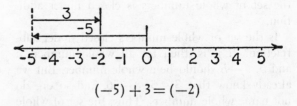

$$(-5) + 3 = (-2)$$

Finally, let us find the sum when both of the addends are negative integers, such as $(-3) + (-2) = (-5)$.

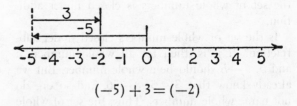

$$(-3) + (-2) = (-5)$$

These number lines give us an idea about adding integers. However, we certainly do not want to draw a number line every time we have to add integers. Furthermore, the number lines merely show how certain sums are found and do not necessarily prove how to add in all cases. We should like to develop the addition of integers according to the properties we have found for addition of whole numbers and without having to use a number line.

ADDITIVE INVERSES

To invent the negative integers we assumed that for every nonzero whole number a there exists the *negative* of a, denoted by $-a$, such that $a + (-a) = 0$.

Definition 4–3:

In the statement $a + (-a) = 0$, a and $-a$ are called *additive inverses* of each other.

This means that a is the additive inverse of $-a$ and also that $-a$ is the additive inverse of a. The term "additive inverse" certainly stems from the use of addition and the fact that $a + (-a)$ names the identity number of addition.

ADDITION OF INTEGERS

Definition 4–2 merely invents the integers. In order for the integers to be useful we must know how to calculate with them. We want the fundamental properties of the whole numbers and the properties of the operations on them to also be true for the set integers. In order to discover definitions for the various operations on the integers, let us assume that addition and multiplication of integers obey the same properties that addition and multiplication of whole numbers obey. In other words, let us assume the following for all integers a, b, and c.

1. *Commutative properties:*
$$a + b = b + a$$
$$a \times b = b \times a$$

2. *Associative properties:*
$$(a + b) + c = a + (b + c)$$
$$(a \times b) \times c = a \times (b \times c)$$

3. *Distributive property:*
$$a \times (b + c) = (a \times b) + (a \times c)$$
$$(b + c) \times a = (b \times a) + (c \times a)$$

4. *Identity numbers:*
$$a + 0 = a = 0 + a$$
$$a \times 1 = a = 1 \times a$$

5. The set of integers is closed under addition, multiplication, and subtraction.

Suppose we are to find the sum of 8 and -2. We might observe a pattern as we know it from the addition of whole numbers.

$$8 + 3 = 11$$
$$8 + 2 = 10$$
$$8 + 1 = 9$$
$$8 + 0 = 8$$
$$8 + (-1) = ?$$
$$8 + (-2) = ?$$

We notice that as we add one less each time, the sum decreases by one each time. If this pattern continues, then $8 + (-1)$ should be 7 and $8 + (-2)$ should be 6.

For a more logical justification of this sum, study the following.

$$8 + (-2) = (6 + 2) + (-2) \quad \text{Rename 8}$$
$$= 6 + [2 + (-2)] \quad \text{Assoc. prop. } +$$
$$\text{Additive inverses}$$
$$= 6 + 0$$
$$\text{Identity number } +$$
$$= 6$$

Exercises 4–2:

Find each sum by using the method illustrated above.

1. $7 + (-3)$ 4. $26 + (-12)$ 7. $(-8) + 11$
2. $12 + (-4)$ 5. $46 + (-24)$ 8. $(-5) + 17$
3. $13 + (-8)$ 6. $87 + (-52)$ 9. $(-9) + 32$

Now let us investigate the case where both addends are negative integers, such as in $(-2) + (-5)$. Since the set of integers is closed under addition, we know that this sum is an integer.

Again, we might observe a pattern.

$$2 + (-5) = (-3)$$
$$1 + (-5) = (-4)$$
$$0 + (-5) = (-5)$$
$$(-1) + (-5) = ?$$
$$(-2) + (-5) = ?$$

If this pattern continues, then $(-1) + (-5)$ should be (-6) and $(-2) + (-5)$ should be (-7).

In deriving the sum of (-2) and (-5), let us consider the following.

$7 + [(-2) + (-5)]$
$= (5 + 2) + [(-2) + (-5)]$　　Rename 7

　　　　　　　　　　　　　　Comm. and
　　　　　　　　　　　　　　assoc. prop. +

$= [5 + (-5)] + [2 + (-2)]$

　　　　　　　　　　　　　　Additive inverses

$= 0$　　　　　　$+ 0$

　　　　　　　　　　　　　　Addition

$= 0$

Since $7 + [(-2) + (-5)] = 0$, then $(-2) + (-5)$ must be the additive inverse of 7. That is, $(-2) + (-5) = (-7)$.

Exercises 4–3:

Find each sum.

1. $(-2) + (-3)$　　6. $(-21) + (-14)$
2. $(-7) + (-5)$　　7. $(-35) + (-22)$
3. $(-8) + (-4)$　　8. $(-18) + (-41)$
4. $(-10) + (-3)$　　9. $(-72) + (-56)$
5. $(-2) + (-17)$　　10. $(-105) + (-17)$

Finally, let us consider a sum of the type $8 + (-13)$.

　　　　　　　　　　　　　　Rename -13
$8 + (-13) = 8 + [(-8) + (-5)]$

　　　　　　　　　　　　　　Assoc. prop. +
$= [8 + (-8)] + (-5)$

　　　　　　　　　　　　　　Additive inverses
$= 0$　　　　$+ (-5)$

　　　　　　　　　　　　　　Identity number +
$= -5$

Exercises 4–4:

Find each sum.

1. $4 + (-7)$　　8. $7 + (-5)$
2. $(-14) + 5$　　9. $(-7) + (-5)$
3. $(-21) + 15$　　10. $121 + (-17)$
4. $8 + (-11)$　　11. $82 + 75$
5. $52 + (-31)$　　12. $(-17) + (-61)$
6. $7 + 5$　　13. $(-44) + 27$
7. $(-7) + 5$　　14. $(-44) + (-27)$

SUBTRACTION OF INTEGERS

We already noticed that the set of whole numbers is not closed under subtraction. For example, $7 - 15$ and $3 - 9$ do not name whole numbers. An important reason for inventing the set of integers is that we desire a set of numbers that is closed under subtraction.

Let us examine some subtractions in the set of whole numbers and some additions in the set of integers.

$$5 - 3 = 2 \qquad 5 + (-3) = 2$$
$$15 - 8 = 7 \qquad 15 + (-8) = 7$$
$$8 - 5 = 3 \qquad 8 + (-5) = 3$$

Apparently there is a close connection between each addition and subtraction since the results are the same. Also, except for some of the signs, the same numerals are involved.

We still want addition and subtraction to be inverse operations for the set of integers. Recall that the difference between the whole numbers 5 and 3, denoted by $5 - 3$, is some number that when added to 3 yields a sum of 5.

$$3 + (5 - 3) = 5$$

Suppose we are to find the difference of two integers, such as $7 - (-3)$. Using our previous interpretation of subtraction, $7 - (-3)$ names some number, call it n, such that when added to -3 yields a sum of 7.

$$n + (-3) = 7$$

Since we know how to add integers, the value of n is easily determined.

$$n + (-3) = 7$$
$$10 + (-3) = 7$$

Hence, $7 - (-3) = 10$ because $10 + (-3) = 7$. Other such examples are:

$$8 - (-5) = 13 \text{ because } 13 + (-5) = 8$$
$$3 - (-4) = 7 \text{ because } 7 + (-4) = 3$$

Now compare the following examples to those above.

$$8 + 5 = 13 \text{ because } 13 - 5 = 8$$
$$3 + 4 = 7 \text{ because } 7 - 4 = 3$$

The result of *subtracting* -5 from 8 is the same as *adding* 5 to 8. The result of *adding* -5

The Set of Integers

to 13 is the same as *subtracting* 5 from 13. Similar relationships can be seen in the other examples. These relationships lead us to the following definition of subtraction of integers.

Definition 4–4:

The *difference* between any two integers *a* and *b*, denoted by $a - b$, is the integer $a + (-b)$.

In other words, *to subtract an integer* we can *add the opposite (additive inverse) of the integer.* For all integers *a* and *b*, $a - b = a + (-b)$. For example:

$$5 - 7 = 5 + (-7) = -2$$
$$13 - (-8) = 13 + 8 = 21$$
$$(-15) - 7 = (-15) + (-7) = -22$$
$$(-9) - (-6) = (-9) + 6 = -3$$

Exercises 4–5:

Find each difference.

1. $7 - (-4)$
2. $(-8) - 12$
3. $11 - 7$
4. $8 - 17$
5. $(-13) - 8$
6. $(-13) - (-8)$
7. $13 - 8$
8. $13 - (-8)$
9. $(-14) - 17$
10. $72 - 185$
11. $72 - (-185)$
12. $0 - 12$

MULTIPLICATION ON A NUMBER LINE

As stated earlier, we desire multiplication of integers to obey the same properties that multiplication of whole numbers obeys. Before using these properties in discovering how to multiply integers, let us use the number line to gain some understanding of multiplication of integers.

In case both factors are positive integers, there is no problem since this is the same as multiplying two whole numbers.

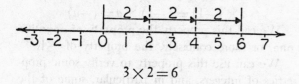

$$3 \times 2 = 6$$

Since multiplication is commutative, $3 \times (-2) = (-2) \times 3$. We need only investigate one of these products. It is difficult to give meaning

to $(-2) \times 3$ in terms of addition since using 3 as an addend -2 times is somewhat elusive. So let us show $3 \times (-2)$ as $(-2) + (-2) + (-2)$ on a number line.

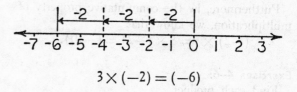

$$3 \times (-2) = (-6)$$

MULTIPLICATION OF INTEGERS

Now let us use the commutative and associative properties of multiplication, the distributive property of multiplication over addition, and additive inverses to establish rules for multiplying integers.

We need not consider the case where both factors are positive integers since this is the same as both factors being whole numbers. Let us begin by considering the case where either factor is a negative integer.

Study the following pattern.

$$5 \times 3 = 15$$
$$5 \times 2 = 10$$
$$5 \times 1 = 5$$
$$5 \times 0 = 0$$
$$5 \times (-1) = ?$$
$$5 \times (-2) = ?$$

We notice that as the second factor decreases by one each time, the product decreases by 5 each time. If this pattern continues, then $5 \times (-1) = (-5)$ and $5 \times (-2) = (-10)$.

In deriving the product
$$5 \times (-2)$$
let us consider the following.

$2 + (-2) = 0$	Additive inverses
$5 \times [2 + (-2)] = 0$	Mult. by zero
$(5 \times 2) + [5 \times (-2)] = 0$	Dist. prop.
$10 + [5 \times (-2)] = 0$	Multiplication

Notice that the number $5 \times (-2)$ is added to 10 and the sum is 0. Hence $5 \times (-2)$ must be the additive inverse of 10. That is, $5 \times (-2) = (-10)$.

This is in agreement with our idea of multiplication in terms of repeated addition.

$$5 \times (-2) = (-2) + (-2) + (-2) + (-2) + (-2) = (-10)$$

Furthermore, by the commutative property of multiplication, we know that

$$(-2) \times 5 = 5 \times (-2) = (-10).$$

Exercises 4–6:

Find each product.

1. $4 \times (-3)$	7. $26 \times (-4)$
2. $8 \times (-2)$	8. $21 \times (-13)$
3. $(-4) \times 6$	9. $321 \times (-7)$
4. $(-5) \times 7$	10. $(-5) \times 205$
5. $12 \times (-10)$	11. $(-250) \times 3$
6. $(-12) \times 10$	12. $(-126) \times 23$

Only one more case remains to be investigated —finding the product of two negative integers.

Knowing how to find the product of a positive integer and a negative integer, let us observe the following pattern.

$$3 \times (-4) = (-12)$$
$$2 \times (-4) = (-8)$$
$$1 \times (-4) = (-4)$$
$$0 \times (-4) = 0$$
$$(-1) \times (-4) = ?$$
$$(-2) \times (-4) = ?$$
$$(-3) \times (-4) = ?$$

As the first factor decreases by one each time, the product increases by four each time. If this pattern continues, then $(-1) \times (-4) = 4$, $(-2) \times (-4) = 8$, and $(-3) \times (-4) = 12$.

In deriving the product

$$(-3) \times (-4)$$

let us consider the following.

Additive inverses

$$4 + (-4) = 0$$

Mult. by zero

$$(-3) \times [4 + (-4)] = 0$$

Dist. prop.

$$[(-3) \times 4] + [(-3) \times (-4)] = 0$$

Multiplication

$$-12 + [(-3) \times (-4)] = 0$$

Notice that the number $(-3) \times (-4)$ is added to -12 and the sum is 0. Hence, $(-3) \times$ (-4) must be the additive inverse of -12. That is, $(-3) \times (-4) = 12$.

You have no doubt noticed a very close relationship between the multiplication of integers and the multiplication of whole numbers. The numerical computation is the same, but the sign or signs used to denote negative integers must be observed very closely in order to state the correct product numeral. We can conclude the following about multiplying integers.

1. The product of two positive integers is a positive integer.

2. The product of a positive integer and a negative integer is a negative integer.

3. The product of two negative integers is a positive integer.

Exercises 4–7:

Find each product.

1. $(-3) \times (-5)$	8. $(-12) \times 10$
2. $(-7) \times (-4)$	9. 12×10
3. $(-7) \times 4$	10. $15 \times (-7)$
4. $7 \times (-4)$	11. $122 \times (-3)$
5. 7×4	12. $(-122) \times (-3)$
6. $(-12) \times (-10)$	13. $402 \times (-4)$
7. $12 \times (-10)$	14. $(-315) \times (-2)$

PROPERTY OF NEGATIVE ONE

Study the following.

$$(-1) \times 7 = (-7) \qquad (-1) \times (-7) = 7$$
$$15 \times (-1) = (-15) \qquad (-13) \times (-1) = 13$$

We know that the number one is the identity number of multiplication. But what happens when we multiply by -1? Instead of the product being identical to the other factor, the product is the negative (or opposite) of the other factor.

$$a \times 1 = 1 \times a = a$$
$$a \times (-1) = (-1) \times a = (-a)$$

We call this property the **property of negative one,** or more concisely, the property of -1.

We can use this property to verify some properties of integers, and in particular, some of the notions we already have about multiplication of integers. In the following examples, a, b, and c represent integers.

DIVISION OF INTEGERS

Every division can be stated as a multiplication by using the idea of multiplication and division as inverse operations. Recall that division of whole numbers was defined so that if

$a \times b = c$ and $b \neq 0$, then $a = c \div b$.

Then division of whole numbers can be defined so that if

$c \div b = a$ and $b \neq 0$, then $a \times b = c$.

Restating each of the following divisions of whole numbers is also the same as restating the division of two positive integers.

$$20 \div 4 = 5 \quad \text{because} \quad 5 \times 4 = 20$$
$$18 \div 3 = 6 \quad \text{because} \quad 6 \times 3 = 18$$
$$250 \div 25 = 10 \quad \text{because} \quad 10 \times 25 = 250$$

Now let us investigate the case where the divisor is a negative integer.

$$24 \div (-4) = q \text{ so } q \times (-4) = 24$$

Then we ask ourselves, "What number multiplied by (-4) yields a product of 24?" Obviously, $q = (-6)$.

Since $(-6) \times (-4) = 24$, then $24 \div (-4) = (-6)$.

Of course, we also know that $24 \div 4 = 6$, so again we see that the important thing to learn about division of integers is observing the negative signs closely.

Suppose the dividend is a negative integer and the divisor is a positive integer, as in the following example.

$$(-32) \div 8 = q \text{ so } q \times 8 = (-32)$$

Ask yourself, "What number multiplied by 8 yields a product of -32?" From knowing how to multiply integers we determine that $q = (-4)$.

Since $(-4) \times 8 = (-32)$, then $(-32) \div 8 = (-4)$.

Thus far, we can conclude the following about division of integers.

1. The quotient of two positive integers is a positive number. (It may or may not be an integer since the set of integers is not closed under division.)

2. The quotient of a positive integer and a negative integer is a negative number. (It may or may not be an integer.)

Prop. of −1
$$(-a) \times b = [(-1) \times a] \times b$$
Assoc. prop. ×
$$= (-1) \times (a \times b)$$
Prop. of −1
$$= -(a \times b)$$

Prop. of −1
$$-(a+b) = (-1) \times (a+b)$$
Dist. prop.
$$= [(-1) \times a] + [(-1) \times b]$$
Prop. of −1
$$= (-a) + (-b)$$

Prop. of −1
$$(-a) \times (-b) = [(-1) \times a] \times [(-1) \times b]$$
Comm. and assoc. prop. ×
$$= [(-1) \times (-1)] \times (a \times b)$$
Multiplication
$$= 1 \times (a \times b)$$
Identity number ×
$$= a \times b$$

Def. of subtraction
$$a \times (b-c) = a \times [b + (-c)]$$
Dist. prop.
$$= (a \times b) + [a \times (-c)]$$
Multiplication
$$= (a \times b) + [-(a \times c)]$$
Def. of subtraction
$$= (a \times b) - (a \times c)$$

Exercises 4–8:
Find each product.

1. $0 \times (-52)$
2. $(-18) \times (-3)$
3. $82 \times (-13)$
4. 126×0
5. $315 \times (-1)$
6. $(-14) \times 0$
7. $(-1) \times 624$
8. $(-36) \times 3$
9. $(-12) \times (-12)$
10. $(-150) \times 6$
11. $(-15) \times (-6)$
12. $(-1) \times (-1)$

Exercises 4–9:

Find each quotient.

1. $10 \div (-2)$
2. $(-16) \div 8$
3. $72 \div (-9)$
4. $45 \div (-5)$
5. $(-63) \div 7$
6. $81 \div (-9)$
7. $75 \div 15$
8. $100 \div 4$
9. $100 \div (-4)$
10. $(-100) \div 4$
11. $256 \div (-8)$
12. $(-168) \div 14$

It may be the case that both the dividend and the divisor are negative integers.

$$(-18) \div (-3) = q \text{ so } q \times (-3) = (-18)$$

What number when multiplied by (-3) yields a product of -18? Again, by multiplication we know that $q = 6$.

Since $6 \times (-3) = (-18)$, then $(-18) \div (-3) = 6$.

Now we can conclude that the quotient of two negative integers is a positive number. (It may or may not be an integer.)

Exercises 4–10:

1. $(-34) \div (-2)$
2. $(-19) \div (-19)$
3. $(-19) \div 19$
4. $(-21) \div (-3)$
5. $(-56) \div (-8)$
6. $132 \div (-11)$
7. $132 \div 11$
8. $(-132) \div 11$
9. $(-132) \div (-11)$
10. $13 \div (-1)$
11. $(-13) \div (-1)$
12. $99 \div 11$
13. $(-99) \div 11$
14. $(-99) \div (-11)$

CHAPTER FIVE

SOLVING EQUATIONS AND PROBLEMS

RELATION SYMBOLS

By using mathematical symbols, we are constantly building a language. In many respects it is more concise than the English language. Because of its brevity we must have a thorough understanding of the meaning of the symbols used in writing a number sentence. Study the meaning of each symbol given below.

$$= \quad \textit{is equal to}$$
$$\neq \quad \textit{is not equal to}$$
$$< \quad \textit{is less than}$$
$$> \quad \textit{is greater than}$$

These symbols serve as verbs in number sentences. Since they show how two numbers are related, they are called *relation symbols*.

The sentence $5 + 3 = 8$ is true since $5 + 3$ and 8 are two names for the same number. But the sentence $7 - 3 = 8$ is false since $7 - 3$ and 8 do not name the same number.

What relation symbol can we write between $7 - 3$ and 8 so that a true sentence is formed? Since $7 - 3$ is not equal to 8, we see that $7 - 3 \neq 8$ is a true sentence.

The sentence $5 - 2 < 4 + 3$ is true since $5 - 2$ or 3 is less than $4 + 3$ or 7. But the sentence $8 + 5 < 9 - 6$ is false since $8 + 5$ or 13 is greater than $9 - 6$ or 3. Hence, $8 + 5 > 9 - 6$ is a true sentence.

The sentence $17 - 5 < 12$ is false since $17 - 5$ or 12 is equal to 12. Hence, we can change the relation symbol $<$ to the relation symbol $=$ and form the true sentence $17 - 5 = 12$.

Exercises 5–1:

Write T before each true sentence below, and write F before each false sentence.

1. $7 + 6 = 15$
2. $14 < 17 + 3$
3. $8 + 2 = 10$
4. $15 - 5 > 6$
5. $23 > 29 + 1$
6. $7 + 6 < 11 - 3$
7. $6 + 9 = 23 - 8$
8. $9 + (-1) = 3 - (-5)$
9. $32 - 2 = 2 \times 15$
10. $3 \times 4 < 5 \times 2$

GROUPING SYMBOLS

Punctuation marks are used in any language to make clear what we want to say. Study how punctuation marks change the meaning of the following unpunctuated sentence.

Bob said Betty is cute.
Bob said, "Betty is cute."
"Bob," said Betty, "is cute."

Punctuation marks are just as important in number sentences as they are in English sentences.

Study the following expression. What number does it name?

$$7 \times 2 + 5$$

Without being told by a symbol or some other means, we do not know whether to do the multiplication or the addition first.

If we multiply first:

$$7 \times 2 + 5 = 14 + 5 = 19$$

If we add first:

$$7 \times 2 + 5 = 7 \times 7 = 49$$

To avoid the confusion of such an expression naming two different numbers, let us use parentheses ( ) to indicate which operation is to be done first.

$$(7 \times 2) + 5 = 14 + 5 = 19$$
$$7 \times (2 + 5) = 7 \times 7 = 49$$

When part of a number sentence is enclosed within parentheses, think of that part as naming but one number. Think of (7×2) as naming 14 and think of $(2 + 5)$ as naming 7.

It is commonly agreed that when more than one operation, or all of the operations, are indicated in the same expression, we multiply and divide first, then add and subtract.

$$5 + 3 \times 4 \text{ means } 5 + (3 \times 4)$$
$$7 - 6 \div 2 \text{ means } 7 - (6 \div 2)$$
$$7 + 3 \times 8 - 5 \text{ means } 7 + (3 \times 8) - 5$$

In case only addition and subtraction are indicated in an expression (or only multiplication and division), we will perform the operations in the order indicated from left to right.

$$8 + 6 - 9 \text{ means } (8 + 6) - 9$$
$$6 \times 4 \div 3 \text{ means } (6 \times 4) \div 3$$

It is not always necessary to write the multiplication sign. Multiplication is indicated in each of the following.

$$6(5 + 4) \text{ means } 6 \times (5 + 4)$$
$$9(15) \text{ means } 9 \times 15$$
$$12n \text{ means } 12 \times n$$
$$(n + 2)(7 - 5) \text{ means } (n + 2) \times (7 - 5)$$

Study how the parentheses are handled in the following sentences.

$$(16 \div 2)(7 + 6) = 8 \times 13 = 104$$
$$(13 - 4) + (12 \div 3) = 9 + 4 = 13$$

Sometimes the sentence becomes so complicated that we need more than one set of parentheses. Instead of two sets of parentheses, let us use brackets [ ] for the second set. In such cases, we handle the innermost groupings first.

$$[4 \times (3 + 2)] - 8 = [4 \times 5] - 8$$
$$= 20 - 8$$
$$= 12$$
$$60 + [(8 \div 2) \times (4 + 3)] = 60 + [4 \times 7]$$
$$= 60 + 28$$
$$= 88$$

Exercises 5–2:

What number is named by each of the following?

1. $(12 - 8) \times 7$ 6. $[6 + (7 - 2)] + 8$
2. $9 \times (32 \div 8)$ 7. $40 \div [(18 \div 3) + 2]$
3. $(11 - 6) + 12$ 8. $[4 + (7 - 2)](9 - 3)$
4. $(4 \times 2)(20 \div 5)$ 9. $[14 \div (6 + 1)] \div 2$
5. $28 \div (10 - 3)$ 10. $8 - [7(5 - 3) - 6]$

Write parentheses, brackets, or both parentheses and brackets in each of the following expressions so that it will name the number indicated after it.

11. $30 - 12 \div 3 \times 2$ Number: 3
12. $30 - 12 \div 3 \times 2$ Number: 52
13. $30 - 12 \div 3 \times 2$ Number: 12
14. $30 - 12 \div 3 \times 2$ Number: 28
15. $30 - 12 \div 3 \times 2$ Number: 22

NUMBER SENTENCES

The symbols used in writing a number sentence are members of one of the following sets.

Number symbols or numerals:
   0, ⅔, 4.7, 5, 72, 119, . . .

Operation symbols:
   $+, -, \times, \div, \ldots$

Relation symbols:
   $=, \neq, <, >, \ldots$

Grouping symbols:
   $(\ ), [\ ], \ldots$

Placeholder symbols or variables:
   $\square, n, x, y, t, \ldots$

To write a number sentence we write a relation symbol between two different combinations of the other symbols.

A very important property of a number sentence which does not contain a placeholder symbol is that it is either true or false, but not both. For example, the following number sentences are classified as true or false.

True	*False*
$4 + 7 = 19 - 8$	$6 + 9 = 17 - 8$
$5 \times 3 < 14 + 6$	$27 - 5 < 2 \times 9$
$48 \div 16 \neq 5$	$5 + 4 > 5 \times 4$

Write T before each true sentence below and write F before each false sentence.

1. $5(3+4) = 21 + 14$
2. $(8 \div 4) + 7 < 4 \times 6$
3. $3 \times 4 > 20 - 8$
4. $6 + 8 \neq 15 \div 3$
5. $3(4+2) = (3 \times 4) + 2$
6. $(3+4)\ (5-5) < 5$

Before each sentence below write T if it is true, F if it is false, and O if it is open.

1. $(5+3) \div k = 2$
2. $4 - (-6) = 10$
3. $4y + (18 \div 6) < 4$
4. $3 + (-8) = 17 - (2 \times 10)$
5. $72 > (24 \div x) + 56$
6. $(16 \div 4) + 13 < (3 \times 4) + 5$

OPEN SENTENCES

In previous lessons we determined whether certain sentences were true or false. Now let us examine the following sentences to learn more about when a number sentence is true, when it is false, and when we are unable to determine which it is.

1. Harry Truman was elected President of the United States.
2. Julius Zulk was elected President of the United States.
3. He was elected President of the United States.

You know that sentence 1 is true, and a little checking of history will show that sentence 2 is false. But what about sentence 3?

Until the word *he* is replaced by the name of a person, we are unable to tell whether sentence 3 is true or false.

Now consider the following number sentences.

$$7 + 6 = 13$$
$$9 - 5 = 27$$
$$8 + n = 15$$

Certainly, $7 + 6 = 13$ is true and $9 - 5 = 27$ is false, but we cannot tell whether $8 + n = 15$ is true or false until n is replaced by a numeral.

Definition 5–1:
Mathematical sentences that contain letters (or some other symbols) to be replaced by numerals, and are neither true nor false, are called *open sentences.*

Examples of open sentences are given below.

$$n - 15 = 7 \qquad 3x + 2 < 12$$
$$(-14)\, t = 64 \qquad 26(y - 3) > 47$$

REPLACEMENT SET

Consider the replacements for the pronoun *she* in the following sentence.

She was a great musician.

Certainly it would be sensible to replace *she* with the name of a person. It would not be sensible to replace *she* with the name of a state, a building, a ship, and so on. If we want the resulting sentence to be meaningful, we must know the set from which we can select the replacements for the pronoun *she*.

This idea also underlies an open sentence. We must know which set of numbers we are allowed to use when replacing a placeholder symbol or variable with a numeral.

Definition 5–2:
The set of numbers whose names are to be used as replacements for a variable is called the *replacement set.*

Solution Set

Use $\{1,2,3,4,5,6\}$ as the replacement set for x in the following open sentence.

$$3 + x < 7$$

We can replace x by each of the numerals 1, 2, 3, 4, 5, and 6 to determine which of them make the resulting sentence true.

True	False
$3 + 1 < 7$	$3 + 4 < 7$
$3 + 2 < 7$	$3 + 5 < 7$
$3 + 3 < 7$	$3 + 6 < 7$

We see that only the numerals 1, 2, and 3 can replace x to make the resulting sentence true. Hence, $\{1,2,3\}$ is called the *solution set* for $3 + x < 7$.

Definition 5–3:

The set of replacements for the variable in an open sentence that make the resulting sentence true is called the *solution set*. Each member of the solution set is called a *solution* or a *root* of the open sentence.

In the previous example, {1,2,3} is the solution set, and the roots of the open sentence are 1, 2, and 3.

Study the following examples of finding a solution set.

Example 1: Find the solution set of $n + 8 = 17$ if the replacement set is the set of whole numbers. We know that $9 + 8 = 17$, so 9 is a solution and belongs in the solution set. If n represents any whole number other than 9, the sentence is false. Hence, the only root is 9 and the solution set is {9}.

Example 2: Find the solution set of $7 - y = 10$ if the replacement set is the set of whole numbers. Since 0 is the least of the whole numbers and $7 - 0 = 7$, we know that y must represent a number less than zero. There is no such whole number, so the solution set is Ø. This does not mean that there is no solution set. It merely means that there is no solution in the set of whole numbers.

Example 3: Find the solution set of $7 - y = 10$ if the replacement set is the set of integers. Since $7 - (-3) = 10$, we know that -3 is a root. We could try other integer replacements for y to convince ourselves that this is the only root. Hence, the solution set is {−3}.

Exercises 5–5:

Using {−5,−4,−3,−2,−1,0,1,2,3,4,5,} as the replacement set, find the solution set of each of these open sentences.

1. $n + 7 = 9$
2. $3x + 2 = -7$
3. $3 + k < 1$
4. $7 - r < 0$
5. $(n \div 2) + 3 = 5$
6. $4t - 7 < 10$
7. $21 - n = 21$
8. $13 \times k = -13$
9. $3(2 + n) = 0$
10. $4(15 \div n) = -12$

EQUATIONS

Those number sentences that state that two expressions are names for the same number are called *equations*. The following are examples of an equation.

$$7 + 6 = 13 \quad n = 9$$
$$5 + 2 = 76 \quad 4(5 + r) = 28$$

We see that an equation might be true, such as $7 + 6 = 13$, or it might be false, such as $5 + 2 = 76$, or it might be an open sentence.

If an equation is not an open sentence, all we need do is determine whether it is true or false. Hence, we are primarily concerned with those equations that are open sentences. When no confusion is possible, let us refer to such open sentences as equations.

Definition 5–4:

To *solve an equation* means to find its solution set.

Throughout this chapter consider the replacement set to be the set of integers for all equations, unless directed otherwise.

In an equation such as $n = 5$ or $k = -56$, the solution set is obvious. We can guess to find the solution set of an equation such as $n + 3 = 5$ or $5t = 15$.

Study how the following equations might be solved mentally.

Solve $3n + 5 = 11$.

By previous agreement, the replacement set for n is the set of integers. Ask yourself: What number plus 5 gives a sum of 11? Since $6 + 5 = 11$, then $3n = 6$. Then ask yourself: 3 times what number gives a product of 6? Since $3 \times 2 = 6$, then $n = 2$. The solution set is {2}. Check your answer by replacing n by 2 in the original equation $(3n + 5 = 11)$ and compute to see if this replacement makes the resulting sentence true.

Exercises 5–6:

Solve these equations.

1. $n + 5 = 12$
2. $5k = 35$
3. $36 \div y = 4$
4. $16 + x = 10$
5. $7 + 3t = 19$
6. $4r + 20 = 0$
7. $x(12 - 8) = 28$
8. $8k \div 2 = -16$

ADDITION PROPERTY OF EQUATIONS

As we attempt to solve more complicated equations, we become aware that a more systematic or logical procedure is needed. That is, we should like to develop a sequence of reasoning for restating an equation until it becomes simple enough for us to solve mentally. However, each step in the reasoning process should be based on the properties of numbers, the properties of the operations, or on the properties of equations that we shall assume.

Suppose we are to solve $r + 5 = 17$. We already know that $r = 12$ since $12 + 5 = 17$, but let us examine more closely how to arrive at such a root. Can we somehow restate the equation so that only r remains on one side of the equal sign?

Let us begin by thinking of $r + 5$. We can undo the adding of 5 by adding its additive inverse, which is -5.

$$(r + 5) + (-5) = r + [5 + (-5)]$$
Assoc. prop. $+$
$$= r + 0$$
Additive inverses
$$= r$$
Identity number $+$

Since $r + 5 = 17$ means that $r + 5$ and 17 are two names for the same number, we must also add -5 to 17 in order that the two new expressions still name the same number. Hence, we could solve the equation as illustrated below.

$$r + 5 = 17$$
$$(r + 5) + (-5) = 17 + (-5)$$
Assoc. prop. $+$
$$r + [5 + (-5)] = 12$$
Additive inverses
$$r + 0 = 12$$
Identity number $+$
$$r = 12$$

The first step in the above solution uses what we call the **addition property of equations.**

For all integers a, b, and c,

if $a = b$, then $a + c = b + c$.

In other words, we can add the same number to both sides of an equation. Of course, we know that adding a negative integer is equivalent to subtracting its opposite. Hence, we do not need such a property for subtraction.

Study the following examples.

Example 1: Solve $17 = n - 8$.

Add. prop. of equations
$$17 + 8 = (n - 8) + 8$$
Addition
$$25 = [n + (-8)] + 8$$
Assoc. prop. $+$
$$25 = n + [(-8) + 8]$$
Additive inverses
$$25 = n + 0$$
Identity number $+$
$$25 = n$$

Example 2: Solve $12 + x = 31$.

Since addition is commutative, we can add -12 on the right or on the left of $12 + x$ and 31.
$$12 + x = 31$$
Add. prop. of equations
$$(-12) + (12 + x) = (-12) + 31$$
Assoc. prop. $+$
$$[(-12) + 12] + x = 19$$
Additive inverses
$$0 + x = 19$$
Identity number $+$
$$x = 19$$

Exercises 5–7:

Find the solution set for each equation.

1. $k + 7 = 21$
2. $29 + x = 36$
3. $y - 12 = 43$
4. $8 + r = -9$
5. $-5 + n = 72$
6. $t + 6 = -18$
7. $17 = r + 6$
8. $n - (-3) = 11$
9. $-7 = r + 15$
10. $76 + n = 76$

MULTIPLICATION PROPERTY OF EQUATIONS

We can denote division by several symbols. In an equation such as $n \div 3 = 17$ it is convenient to state $n \div 3$ as $\dfrac{n}{3}$.

Again, to restate the equation so that it can be solved mentally, we should like to restate the equation so that only n remains on one side of the equal sign.

We can think of undoing the dividing by 3 by multiplying by 3, since multiplication and division are inverse operations.

$$n \div 3 = 17$$
$$(n \div 3) \times 3 = 17 \times 3$$
$$n = 51$$

Another way of writing this is:

$$\frac{n}{3} = 17$$

$$\frac{n}{3} \times 3 = 17 \times 3$$

$$\frac{n \times 3}{3} = 51$$

$$n \times \frac{3}{3} = 51$$

$$n \times 1 = 51$$

$$n = 51$$

In the first step of this solution we used the **multiplication property of equations.**

For all integers a, b, and c, if $a = b$, then $a \times c = b \times c$.

Study how this property is used in solving the following equation.

$$-14 = \frac{a}{4}$$

$$(-14)(4) = \frac{a}{4} \times 4$$

$$-56 = \frac{a \times 4}{4}$$

$$-56 = a \times \frac{4}{4}$$

$$-56 = a \times 1$$

$$-56 = a$$

Fractions occurred in both of the preceding examples. You are probably familiar with the operations on these fractional numbers. These operations will be fully explained in the chapter dealing with rational numbers.

Exercises 5–8:
Solve the following equations.

1. $\dfrac{c}{5} = 8$ 5. $\dfrac{n}{-4} = -7$ 9. $\dfrac{a}{2+5} = 8$

2. $\dfrac{x}{-3} = 9$ 6. $\dfrac{x}{12} = -6$ 10. $\dfrac{t}{4(-5)} = -8$

3. $21 = \dfrac{n}{10}$ 7. $25 = \dfrac{c}{3-4}$ 11. $3(-6) = \dfrac{n}{5}$

4. $0 = \dfrac{t}{3}$ 8. $\dfrac{a}{-9} = 1$ 12. $\dfrac{r}{-1} = 1$

DIVISION PROPERTY OF EQUATIONS

Consider solving the equation $4n = 32$. In this case n is multiplied by 4. Since multiplication and division are inverse operations, we can undo multiplying by 4 by dividing by 4. So we might divide both $4n$ and 32 by 4, as shown in the following example.

$$4n = 32$$

$$\frac{4n}{4} = \frac{32}{4}$$

$$\frac{4}{4} \times n = 8$$

$$1 \times n = 8$$

$$n = 8$$

In this case we have used the **division property of equations.**

For all integers a, b, and c, where $c \neq 0$, if $a = b$, then $\dfrac{a}{c} = \dfrac{b}{c}$.

We might ask why we have a division property and no subtraction property. Once we have invented the set of rational (fractional) numbers, we can show that the division property is no longer needed. Until that time we will use this property of equations.

Study how this property is used in the following examples.

Example 1: Solve $5n = 35$.

$$5n = 35$$

$$\frac{5n}{5} = \frac{35}{5}$$ Div. prop. of equations

$$\frac{5}{5} \times n = 7$$

$$1 \times n = 7$$

$$n = 7$$

Example 2: Solve $-8c = 56$.

$$-8c = 56$$

$$\frac{-8c}{-8} = \frac{56}{-8}$$ Div. prop. of equations

$$\frac{-8}{-8} \times c = -7$$

$$1 \times c = -7$$

$$c = -7$$

Exercises 5–9:

Solve these equations.

1. $4n = 36$
2. $42 = -6r$
3. $-7k = 63$
4. $-5x = -75$
5. $-k = 47$
6. $72 = 8x$
7. $-52 = -a$
8. $t(5 + 2) = 91$
9. $n(5 - 9) = -24$
10. $x(21 \times 3) = 0$

SOLVING EQUATIONS

As we attempt to solve more and more complicated equations, we may need to use more than one of the properties of equations or use the same property more than once.

In solving an equation such as $5t + 6 = 21$ it is generally advisable to use the property that is most convenient for changing the expression $5t + 6$ to $5t$ first. Then use the property for changing $5t$ to t. Study the following example.

$$5t + 6 = 21$$

Addition property of equations

$$(5t + 6) + (-6) = 21 + (-6)$$

Assoc. prop. $+$

$$5t + [6 + (-6)] = 21 + (-6)$$

Addition

$$5t + 0 = 15$$

Identity number $+$

$$5t = 15$$

Division property of equations

$$\frac{5t}{5} = \frac{15}{5}$$

Division

$$t = 3$$

Then we can check our work by replacing t by 3 in the original equation.

$$5t + 6 = 21$$
$$5(3) + 6 = 21$$
$$15 + 6 = 21$$
$$21 = 21$$

Since we have shown that $5(3) + 6$ and 21 name the same number, we know that 3 is a root of the equation.

Since we know that adding a number and its additive inverse is equivalent to adding zero, we can combine some of the steps when writing the previous solution. Also, we may make some of the calculations mentally in the process. However, the example shows the thinking steps necessary in solving the equation.

As you study the following examples, think of the reason or reasons for each step.

Example 1: Solve $\frac{n}{4} - 13 = 3$.

$$\frac{n}{4} - 13 = 3$$

$$\left[\frac{n}{4} + (-13)\right] + 13 = 3 + 13$$

$$\frac{n}{4} = 16$$

$$\frac{n}{4} \times 4 = 16 \times 4$$

$$n = 64$$

Example 2: Solve $\dfrac{x-5}{4} = 6.$

$$\frac{x-5}{4} = 6$$

$$\frac{x-5}{4} \times 4 = 6 \times 4$$

$$x - 5 = 24$$

$$(x-5) + 5 = 24 + 5$$

$$x = 29$$

Example 3: Solve $\dfrac{3t+7}{2} = 26.$

$$\frac{3t+7}{2} = 26$$

$$\frac{3t+7}{2} \times 2 = 26 \times 2$$

$$3t + 7 = 52$$

$$(3t+7) + (-7) = 52 + (-7)$$

$$3t = 45$$

$$\frac{3t}{3} = \frac{45}{3}$$

$$t = 15$$

Example 4: Solve $5(2t-14) = 60.$

Solution 1:
$$5(2t-14) = 60$$

$$\frac{5(2t-14)}{5} = \frac{60}{5}$$

$$2t - 14 = 12$$

$$(2t-14) + 14 = 12 + 14$$

$$2t = 26$$

$$\frac{2t}{2} = \frac{26}{2}$$

$$t = 13$$

Solution 2:
$$5(2t-14) = 60$$

$$10t - 70 = 60$$

$$(10t - 70) + 70 = 60 + 70$$

$$10t = 130$$

$$\frac{10t}{10} = \frac{130}{10}$$

$$t = 13$$

Exercises 5–10:

Solve these equations. Check your answers by replacing the variable in the original equation with the root you have found.

1. $3n + 4 = 19$

2. $14 = \dfrac{a}{2} - 6$

3. $-26 = 1 - 3x$

4. $15 + 3t = 0$

5. $3t + 5 = 29$

6. $\dfrac{r}{4} + 8 = 7$

7. $4(k + 3) = 0$

8. $\dfrac{3n}{4} = 9$

9. $\dfrac{2c}{5} + 6 = 10$

10. $\dfrac{n+7}{5} = 16$

11. $\dfrac{5n}{3} - 13 = 2$

12. $12 = 11t - 10$

13. $4 = \dfrac{3n-4}{8}$

14. $7t - 18 = 73$

15. $3(4n - 1) = 21$

16. $\dfrac{n+14}{3} = 20$

MORE ABOUT SOLVING EQUATIONS

If a variable occurs more than once in the same equation, it must be replaced by the same numeral in both instances. For example, if either of the letters k in $5k + 2 = 7 + k$ is replaced by the numeral 3, then the other must also be replaced by 3.

Before solving equations where the same variable occurs more than once, let us investigate some expressions of this type.

What is a simpler name for $2y + 5y$? Our first guess would probably be $7y$. We could test our guess for a few replacements of y to see if $2y + 5y = 7y$. For example:

If we replace y by 3, then
$2y + 5y = 2(3) + 5(3) = 6 + 15 = 21$
$7y = 7(3) = 21$
If we place y by -5, then
$2y + 5y = 2(-5) + 5(-5) = -10 + (-25)$
    $= -35$
$7y = 7(-5) = -35$

At least for these two replacements of y the sentence $2y + 5y = 7y$ is true. Since it is impossible to test this equation for all values of y, let us use the properties of numbers and their operations to verify that $2y + 5y = 7y$.

$$
\begin{aligned}
2y + 5y &= (2 \times y) + (5 \times y) \\
&= (2 + 5) \times y \quad \text{Dist. prop.} \\
&= 7 \times y \quad\quad\;\; \text{Addition} \\
&= 7y
\end{aligned}
$$

We can also use the distributive property to show that $9x - 5x = 4x$.

$$
\begin{aligned}
9x - 5x &= 9x + (-5x) \\
&= [9 + (-5)]x \\
&= 4x
\end{aligned}
$$

Since a placeholder or variable names a number, we can treat it just as we do a numeral when solving an equation, as shown in the following.

Example 1: Solve $9t = 40 + t$.

$$
\begin{aligned}
9t &= 40 + t \\
9t + (-t) &= (40 + t) + (-t) \\
8t &= 40 \\
\frac{8t}{8} &= \frac{40}{8} \\
t &= 5
\end{aligned}
$$

Example 2: Solve $5(2n - 3) = 21 + n$.

$$
\begin{aligned}
5(2n - 3) &= 21 + n \\
10n - 15 &= 21 + n \\
(10n - 15) + (-n) &= (21 + n) + (-n) \\
(-n) + [10n + (-15)] &= 21 + 0 \\
[(-n) + 10n] + (-15) &= 21 \\
9n + (-15) &= 21 \\
[9n + (-15)] + 15 &= 21 + 15 \\
9n &= 36 \\
\frac{9n}{9} &= \frac{36}{9} \\
n &= 4
\end{aligned}
$$

Exercises 5–11:

Solve these equations.

1. $7k + 8 = 11k$
2. $8n = 14 + n$
3. $4t - 5 = t + 1$
4. $7r + 3r = 130$
5. $17x - 11x = 42$
6. $c + 3(5 + c) = 23$
7. $a + 14 = 5(a - 2)$
8. $2t + 18 = t + 6$
9. $3(2t - 18) = 11 + t$
10. $15 - t = 2(6 + t)$

TRANSLATNG ENGLISH PHRASES

One of the most important skills in problem-solving is the ability to translate a problem stated in the English language into the language of mathematics. That is, we want to write an open sentence that says essentially the same thing as a "story problem."

We know from our study of the English language that sentences may contain phrases. Before translating sentences, let us investigate what we shall call *open phrases*, such as $k + 5$.

Suppose we want to express John's age 6 years ago and we do not know what John's age is now. We might think as follows.

Number of years in John's age now_____ n
Number of years in John's age 6 years ago_____ $n - 6$

Suppose Bob's age is 5 years more than 3 times his sister's age. How can we express Bob's age?

Number of years in his sister's age_____ s
3 times the number of years in his sister's age_____ $3s$
5 years more than 3 times the number of years in his sister's age____ $3s + 5$

In making the translation from English to mathematics, we first choose some letter to use as the variable. Then decide which operation or operations say essentially the same thing as the English words.

Exercises 5–12:

Translate the following English phrases into open phrases. Use the letter n for the variable in each open phrase.

1. Seven more than some number
2. Three less than 2 times some number
3. The sum of a number and twice the number
4. Mary's age 8 years from now
5. The number of cents in n nickels and $(7 - n)$ dimes
6. A man's age is 9 years greater than 2 times his son's age
7. The sum of 3 times some number and 4 times the number
8. Five more than twice the number of dollars Jim has
9. The number of feet in the distance around a square
10. Bob's score on a test if he answered 3 problems incorrectly

TRANSLATING ENGLISH SENTENCES

We usually describe a problem situation in the English language. Some of the English sentences can be translated into mathematical sentences and others cannot be so translated. For example, the sentence "The rose is red" does not lend itself to a mathematical translation.

Consider the following sentence.

John is 14 years old.

This sentence is just as meaningful if stated as follows. Then we can easily translate it into the language of mathematics.

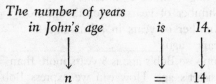

The number of years
in John's age   is   14.

$$n = 14$$

Now consider the following sentence.

Six years ago John was 8 years old.

This sentence is just as meaningful if stated as follows.

The number of years in John's age 6 years ago   was   8.

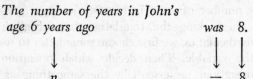

$$n = 8$$

Notice that both of these sentences were re-written in order to emphasize that the variable, in this case *n*, represents a *number*.

Exercises 5–13:

Translate each of these English sentences into open sentences. Use the letter *t* for the variable in each open sentence.

1. If Mark spends 3 dollars he will have 4 dollars left.

2. The product of some number and 12 is 32.

3. The difference between 3 times some number and 8 is 7.

4. Alice received 51 votes, which is 7 more votes than George received.

5. When a certain number is divided by 5 the quotient is −9.

SOLVING PROBLEMS

There is no one set of rules for solving problems, nor is there one way to apply mathematics to the physical world. However, some suggestions can be made for solving problems.

a. Study the problem carefully and think about the situation in terms of which operation or operations to use and what open sentence you might write for the problem.

b. Translate the problem into an open sentence.

c. Solve the open sentence.

d. Use the root or roots of the open sentence to answer the problem.

Study how each of the following problems are translated into an open sentence and how the root of the open sentence is used to answer the problem.

Example 1:

The George Washington Bridge has two end spans of the same length and a center span that is 3600 feet long. The overall length of the bridge is 4800 feet. How long is each end span?

Let $f =$ the number of feet in the length of each end span

$$2f + 3600 = 4800$$
$$(2f + 3600) + (-3600) = 4800 + (-3600)$$
$$2f = 1200$$
$$\frac{2f}{2} = \frac{1200}{2}$$
$$f = 600$$

Each end span is 600 feet long.

Example 2:

Jim and Ed were the only candidates for president. Jim received 52 more votes than Ed. If 264 votes were cast, how many votes did each boy receive?

Let $e =$ the number of votes for Ed
$e + 52 =$ the number of votes for Jim

$$e + (e + 52) = 264$$
$$(e + e) + 52 = 264$$
$$2e + 52 = 264$$
$$(2e + 52) + (-52) = 264 + (-52)$$
$$2e = 212$$
$$\frac{2e}{2} = \frac{212}{2}$$
$$e = 106$$

Ed received 106 votes.
Jim received $e + 52$ or 158 votes.

Example 3:

Jean's age is 7 years more than twice her sister's age. If Jean is 19 years old, how old is her sister?

Let $s =$ the number of years in her sister's age

$$2s + 7 = 19$$
$$(2s + 7) + (-7) = 19 + (-7)$$
$$2s = 12$$
$$s = 6$$

Her sister is 6 years old.

Exercises 5–14:

Solve these problems.

1. The sum of two times a certain number and 6 is 22. What is the number?

2. One number is 3 more than a second number. Their sum is 67. What are the two numbers?

3. Ted weighs 9 pounds less than Roger. Their combined weight is 239 pounds. How much does each boy weigh?

4. A rope 26 feet long is cut into 2 pieces so that one piece is 8 feet longer than the other. How long is the shorter piece of rope?

5. A rectangle is 12 inches long. Its perimeter is 44 inches. How wide is the rectangle?

6. The sum of a number and 4 times the number is 75. What is the number?

7. Rita said, "If I had 40 cents more than twice what I have, I would have $3.30." How much money does Rita have?

8. Robert pays 5 cents each for papers and sells them for 8 cents each. Last week he earned $2.70 by selling papers. How many papers did he sell last week?

9. The difference between a number and 5 times the number is 32. What is the number? (Hint: There are two possible answers—one positive and one negative.)

CHAPTER SIX

RATIONAL NUMBERS

A NEED FOR NEW NUMBERS

Mark read 3 times as many books as Gene. Mark read 8 books. How many books did Gene read?

This problem can easily be translated into the equation $3n = 8$, where n represents the number of books Gene read. If we are allowed to use only the set of integers as the replacement set for n, then $3n = 8$ has no root. This is an indication that the set of integers is not sufficient for all our needs.

If we attempt to solve the above equation, our work may appear as follows:

$$3n = 8$$
$$\frac{3n}{3} = \frac{8}{3}$$
$$n = \frac{8}{3}$$

The symbol $\frac{8}{3}$ means "8 divided by 3." Since the set of integers is not closed under division, we are not surprised that $\frac{8}{3}$ does not name an integer.

The symbol $\frac{8}{3}$ is called a *fraction*. Hence, a fraction is a numeral or a name for a number. In fact, it is a name for a *rational number*.

Definition 6–1:

A *rational number* is any number of the form $\frac{a}{b}$ where a and b are integers and $b \neq 0$. The integer a is called the *numerator* and the integer b is called the *denominator*.

In other words, a rational number is the quotient of two integers where the divisor is not zero. The following fractions name rational numbers.

$$\frac{1}{2}, \frac{4}{3}, \frac{6}{-2}, \frac{0}{5}, \frac{-3}{-1}, \frac{27}{5}, \frac{100}{25}$$

Does it appear that some of these fractions also name integers? In fact, every integer can be named by a fraction.

$$\cdots \frac{-2}{1}, \frac{-3}{1}, \frac{-0}{1}, \frac{1}{1}, \frac{2}{1}, \frac{3}{1}, \cdots$$

Hence, the set of integers is a subset of the set of rational numbers.

Exercises 6–1:

Solve the following equations by using the set of rational numbers as the replacement set.

1. $4x = 17$
2. $-3n = 11$
3. $8t = 9$
4. $7r = 28$
5. $\frac{3a}{5} = 4$

6. $2n + 7 = 12$
7. $5c - 4 = 26$
8. $\frac{5r}{7} = -15$
9. $-3a + 2 = 29$
10. $15 + 4n = -17$

OTHER NAMES FOR RATIONAL NUMBERS

From our knowledge of division of integers we know that $\frac{4}{2} = 2$ and $\frac{6}{2} = 3$. That is, $\frac{4}{2}$ and 2 are two names for the same number and $\frac{6}{2}$ and 3 are two names for the same number. Then ⁵⁄₂ must name a number between 2 and 3.

Since ⁵⁄₂ means $5 \div 2$, we can find another name for ⁵⁄₂ as follows:

$$\frac{5}{2} = 5 \div 2 = (4+1) \div 2$$

$$= (4 \div 2) + (1 \div 2)$$

$$= 2 + \frac{1}{2}$$

A short way to write $2 + \frac{1}{2}$ is 2½. Hence, ⁵⁄₂ = 2½ which means that ⁵⁄₂ and 2½ are two names for the same number. One of these names is as acceptable as the other. However, each has their advantages and disadvantages.

For convenience of reference, let us refer to a numeral such as 2½ as a "mixed numeral."

A person probably gets a better idea of the quantity if we say 5⅜ gallons than if we say ⁴³⁄₈ gallons. However, for most computational purposes the name ⁴³⁄₈ is more convenient.

Exercises 6–2:

Name each of the following by mixed numerals.

1. $\frac{8}{5}$ 4. $\frac{62}{7}$ 7. $\frac{136}{11}$

2. $\frac{41}{10}$ 5. $\frac{38}{9}$ 8. $\frac{236}{25}$

3. $\frac{51}{7}$ 6. $\frac{85}{9}$ 9. $\frac{421}{50}$

THE RATIONAL NUMBER LINE

A number line is a very useful tool for picturing some of our ideas about numbers and operations. Recall that we established a one-to-one correspondence between the set of integers and some selected points on a number line.

Obviously there are many points on the line which are not matched with these numbers. Some of these points (but not all of them) can be matched one-to-one with the rational numbers. Let us consider that segment of the number line from 0 to 1. The point midway between 0 and 1 is matched with the rational number ½.

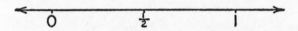

Then we could lay off the segment of length ½ to the left and to the right to locate those points to be matched with the rational numbers . . . , −³⁄₂, −²⁄₂, −½, ⁰⁄₂, ½, ²⁄₂, ³⁄₂, . . .

Those points that separate the segment from 0 to 1 into three parts of equal length, can be matched with the rational numbers ⅓ and ⅔.

Then we could lay off the segment of length ⅓ to the left and to the right to locate those points to be matched with the rational numbers . . . , −⁴⁄₃, −³⁄₃, −²⁄₃, −⅓, 0, ⅓, ⅔, ³⁄₃, ⁴⁄₃, . . .

Similarly, we could locate points to be matched with any rational number. We might be tempted to say at this time that such a one-to-one correspondence would involve every point on the number line, but this is far from true. There still remain many unmatched points on the number line. Later in this chapter we shall discuss those numbers that we match with these points.

MULTIPLICATION OF RATIONAL NUMBERS

Just as we did when we extended the set of whole numbers to the set of integers, we want the operations on the rational numbers to obey the same properties that the operations on the integers obey. In other words, let us assume both commutative properties, both associative properties, the distributive property, 0 as the identity number of addition, and 1 as the identity number of multiplication for the set of rational numbers.

When you first learned about fractions you probably interpreted the fraction ¼ to indicate that some object had been separated in 4 parts of equal size, and you were considering 1 of these parts.

These intuitive ideas also give us a clue to the product of ¼ and ½.

Think of separating something into 4 parts of equal size. Then separate one of these parts into 2 parts of equal size. What part of the object is one of these latter pieces? Obviously ½ of ¼ is ⅛, or ½ × ¼ = ⅛.

By using other fractions, we can find that ½ × ⅕ = ¹⁄₁₀, that ⅓ × ⅕ = ¹⁄₁₅, and so on.

$$\frac{1}{a} \times \frac{1}{n} = \frac{1 \times 1}{a \times n} = \frac{1}{an}$$

for all rational numbers $\frac{1}{a}$ and $\frac{1}{n}$.

Exercises 6–3:

Find each product.

1. $\frac{1}{3} \times \frac{1}{4}$ 4. $\frac{1}{8} \times \frac{1}{5}$ 7. $\frac{1}{12} \times \frac{1}{5}$

2. $\frac{1}{6} \times \frac{1}{2}$ 5. $\frac{1}{2} \times \frac{1}{9}$ 8. $\frac{1}{13} \times \frac{1}{3}$

3. $\frac{1}{7} \times \frac{1}{3}$ 6. $\frac{1}{10} \times \frac{1}{3}$ 9. $\frac{1}{4} \times \frac{1}{17}$

We can get a more general notion of multiplication of rational numbers by investigating the number line.

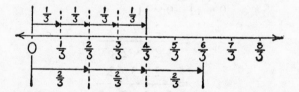

The arrows above the number line show ⅓ + ⅓ + ⅓ + ⅓ = ⁴⁄₃. From our previous idea of multiplication of integers, we may also interpret ⅓ + ⅓ + ⅓ + ⅓ as 4 × ⅓. Then it appears that

$$4 \times \frac{1}{3} = \frac{4 \times 1}{3} = \frac{4}{3}.$$

The arrows below the number line show ⅔ + ⅔ + ⅔ = ⁶⁄₃. In terms of multiplication it appears that

$$3 \times \frac{2}{3} = \frac{3 \times 2}{3} = \frac{6}{3}$$

The first example reveals an important way of renaming any rational number. Let us begin with ⁴⁄₃ and work back through the example.

$$\frac{4}{3} = \frac{4 \times 1}{3} = 4 \times \frac{1}{3}$$

Let us assume that this is true for all rational numbers.

For every rational number $\frac{a}{b}$,

$$\frac{a}{b} = \frac{a \times 1}{b} = a \times \frac{1}{b}.$$

Now let us use these ideas to find the product of ⅔ and ⅘.

$$\frac{2}{3} \times \frac{4}{5} = \left(2 \times \frac{1}{3}\right)\left(4 \times \frac{1}{5}\right)$$

$$= 2 \times \left(\frac{1}{3} \times 4\right) \times \frac{1}{5} \quad \text{Assoc. prop.}$$

$$= 2 \times \left(4 \times \frac{1}{3}\right) \times \frac{1}{5} \quad \text{Comm. prop.}$$

$$= (2 \times 4)\left(\frac{1}{3} \times \frac{1}{5}\right) \quad \text{Assoc. prop.}$$

$$= 8 \times \frac{1}{15}$$

$$= \frac{8 \times 1}{15} = \frac{8}{15}$$

Then we notice that 8 = 2 × 4 and 15 = 3 × 5. We could have multiplied as follows.

$$\frac{2}{3} \times \frac{4}{5} = \frac{2 \times 4}{3 \times 5} = \frac{8}{15}$$

From other such examples, we are led to the following definition of multiplication of rational numbers.

Definition 6–2

For all rational number $\dfrac{a}{n}$ and $\dfrac{r}{s}$,

$$\frac{a}{n} \times \frac{r}{s} = \frac{a \times r}{n \times s} = \frac{ar}{ns}.$$

Exercises 6–4:

Find each product.

1. $\dfrac{3}{7} \times \dfrac{4}{5}$ 5. $\dfrac{5}{13} \times \dfrac{4}{3}$ 9. $\dfrac{5}{11} \times \dfrac{3}{7}$

2. $\dfrac{2}{3} \times \dfrac{5}{7}$ 6. $\dfrac{4}{5} \times \dfrac{2}{9}$ 10. $\dfrac{5}{8} \times \dfrac{3}{11}$

3. $\dfrac{4}{9} \times \dfrac{10}{7}$ 7. $\dfrac{8}{13} \times \dfrac{2}{3}$ 11. $\dfrac{2}{15} \times \dfrac{5}{3}$

4. $\dfrac{0}{7} \times \dfrac{5}{6}$ 8. $\dfrac{1}{2} \times \dfrac{9}{8}$ 12. $\dfrac{3}{8} \times \dfrac{7}{11}$

RENAMING RATIONAL NUMBERS

We have already assumed the number 1 to be the identity number of multiplication. For the set of integers we discovered that $4 \div 4 = 1$, $3 \div 3 = 1$, and so on, for any nonzero number. Since $4 \div 4$ can be written as $\frac{4}{4}$, we can conclude the following.

For any nonzero integer n, the rational number $\dfrac{n}{n} = 1$.

Consider the following multiplication, when 1 is named as a rational number of the form $\dfrac{n}{n}$.

$$\frac{2}{3} \times 1 = \frac{2}{3} \times \frac{2}{2} = \frac{2 \times 2}{3 \times 2} = \frac{4}{6}$$

$$\frac{2}{3} \times 1 = \frac{2}{3} \times \frac{3}{3} = \frac{2 \times 3}{3 \times 3} = \frac{6}{9}$$

Now consider the following.

$$\frac{8}{12} = \frac{4 \times 2}{4 \times 3} = \frac{4}{4} \times \frac{2}{3} = 1 \times \frac{2}{3} = \frac{2}{3}$$

$$\frac{18}{30} = \frac{6 \times 3}{6 \times 5} = \frac{6}{6} \times \frac{3}{5} = 1 \times \frac{3}{5} = \frac{3}{5}$$

From such examples we can make the following assumption.

For any rational number $\dfrac{a}{c}$ and for any nonzero integer r,

$$\frac{a}{c} = \frac{ra}{rc}.$$

In other words, both the numerator and denominator of a rational number can be multiplied or divided by the same nonzero integer without changing the rational number. All we do in the process is write another name for the number. This is the idea behind what is commonly called "reducing fractions" to lowest terms when there is no whole number, other than 1, that will divide both the numerator and the denominator.

Example 1:

$$\frac{4}{5} \times \frac{3}{8} = \frac{4 \times 3}{5 \times 8} = \frac{12}{40} = \frac{4 \times 3}{4 \times 10} = \frac{3}{10}$$

Example 2:

$$\frac{4}{9} \times \frac{3}{14} = \frac{4 \times 3}{9 \times 14}$$

$$= \frac{2 \times 2 \times 3}{3 \times 3 \times 2 \times 7}$$

$$= \frac{2 \times 3 \times 2}{2 \times 3 \times 3 \times 7}$$

$$= \frac{2 \times 3}{2 \times 3} \times \frac{2}{3 \times 7}$$

$$= 1 \times \frac{2}{21}$$

$$= \frac{2}{21}$$

Exercises 6–5:

Find each product. State the answer in the lowest terms.

1. $\dfrac{5}{16} \times \dfrac{4}{5}$ 5. $\dfrac{2}{3} \times \dfrac{3}{4} \times \dfrac{5}{6}$

2. $\dfrac{3}{4} \times \dfrac{8}{15}$ 6. $\dfrac{5}{8} \times \dfrac{4}{7} \times \dfrac{21}{25}$

3. $\dfrac{15}{16} \times \dfrac{20}{21}$ 7. $\dfrac{3}{2} \times \dfrac{4}{9} \times \dfrac{5}{6}$

4. $\dfrac{15}{2} \times \dfrac{8}{5}$ 8. $\dfrac{7}{10} \times \dfrac{5}{8} \times \dfrac{6}{21}$

MIXED NUMERALS IN MULTIPLICATION

One or more of the factors might be named by a mixed numeral. Our previous definition of multiplication covers such cases.

$$5\frac{1}{2} \times \frac{4}{7} = \frac{11}{2} \times \frac{4}{7}$$

$$= \frac{11 \times 4}{2 \times 7}$$

$$= \frac{2 \times 2 \times 11}{2 \times 7}$$

$$= \frac{2}{2} \times \frac{2 \times 11}{7}$$

$$= 1 \times \frac{22}{7}$$

$$= \frac{22}{7}$$

All we need to do is change the mixed numeral to a fraction and proceed as usual.

Exercises 6–6:

Find each product. State the answer in lowest terms.

1. $4\frac{1}{2} \times \frac{4}{5}$ 5. $\frac{6}{7} \times 3\frac{1}{2} \times 4$
2. $5\frac{2}{3} \times \frac{3}{2}$ 6. $4\frac{1}{2} \times \frac{5}{3} \times \frac{4}{5}$
3. $3\frac{1}{5} \times \frac{1}{8}$ 7. $5\frac{1}{2} \times 6\frac{2}{5} \times 3\frac{2}{11}$
4. $4\frac{1}{3} \times 12$ 8. $\frac{3}{2} \times 1\frac{5}{7} \times \frac{7}{8}$

RECIPROCALS OR MULTIPLICATIVE INVERSES

We have seen that the identity number of multiplication of rational numbers is 1. Now we wish to investigate the possibility of two rational numbers having a product of 1.

Study the following examples.

$$\frac{4}{3} \times \frac{3}{4} = \frac{4 \times 3}{3 \times 4} = \frac{12}{12} = 1$$

$$\frac{5}{6} \times \frac{6}{5} = \frac{5 \times 6}{6 \times 5} = \frac{30}{30} = 1$$

In general, for any rational number $\dfrac{r}{s}$ we can find a rational number $\dfrac{s}{r}$ so that

$$\frac{r}{s} \times \frac{s}{r} = 1.$$

Definition 6–3:

If the product of two rational numbers is 1, then the two rational numbers are called the *reciprocals* or the *multiplicative inverses* of each other.

Since $\frac{3}{4} \times \frac{4}{3} = 1$, $\frac{3}{4}$ and $\frac{4}{3}$ are each the reciprocal or the multiplicative inverse of the other. Since 5 can be named as $\frac{5}{1}$, then $\frac{5}{1} \times \frac{1}{5} = 1$ and 5 and $\frac{1}{5}$ are reciprocals of each other.

Does every integer have a reciprocal? Since $-3 \times (-\frac{1}{3}) = 1$, we see that negative integers have reciprocals. What number could n represent so that $0 \times n = 1$? There is no such number n, and we conclude that zero has no reciprocal.

Exercises 6–7:

State the reciprocal of each of the following numbers.

1. $\frac{5}{8}$ 4. $\frac{7}{12}$ 7. $3\frac{1}{2}$
2. 7   5. $\frac{9}{31}$ 8. -9
3. $-\frac{4}{9}$ 6. $-\frac{15}{37}$ 9. $-6\frac{5}{7}$

DIVISION OF RATIONAL NUMBERS

Since multiplication and division are inverse operations, we can state every division as a multiplication. Suppose we are to solve the equation $\frac{5}{7} \div \frac{2}{3} = n$.

$$\frac{5}{7} \div \frac{2}{3} = n$$

$$\frac{5}{7} = n \times \frac{2}{3} \qquad \text{Inverse operations}$$

$$\frac{5}{7} \times 1 = n \times \frac{2}{3} \qquad \text{Identity number} \times$$

$$\frac{5}{7} \times \left(\frac{3}{2} \times \frac{2}{3}\right) = n \times \frac{2}{3} \qquad \begin{array}{l}\text{State 1 as product of}\\ \frac{2}{3} \text{ and its reciprocal.}\end{array}$$

$$\left(\frac{5}{7} \times \frac{3}{2}\right) \times \frac{2}{3} = n \times \frac{2}{3} \qquad \text{Assoc. prop.} \times$$

Since the second factor on both sides of the equation is $\frac{2}{3}$, we conclude that

$$\frac{5}{7} \times \frac{3}{2} = n.$$

Replacing n in the original equation by this expression we have

$$\frac{5}{7} \div \frac{2}{3} = \frac{5}{7} \times \frac{3}{2}.$$

Dividing by a rational number is essentially the same as multiplying by its reciprocal (or multiplicative inverse).

Definition 6–4:

For all rational numbers $\frac{a}{n}$ and $\frac{r}{s}$,

$$\frac{a}{n} \div \frac{r}{s} = \frac{a}{n} \times \frac{s}{r}.$$

Hence, to perform division of rational numbers, merely restate the division as a multiplication and proceed according to the rules for multiplying rational numbers.

Exercises 6–8:

Find each quotient. State your answers in lowest terms.

1. $\dfrac{8}{9} \div \dfrac{4}{9}$ 6. $\dfrac{10}{11} \div \dfrac{2}{5}$ 11. $4\dfrac{1}{5} \div \dfrac{2}{3}$

2. $\dfrac{4}{5} \div \dfrac{8}{15}$ 7. $15 \div 3\dfrac{1}{3}$ 12. $1\dfrac{1}{5} \div 24$

3. $\dfrac{5}{6} \div 4$ 8. $1\dfrac{1}{6} \div \dfrac{1}{9}$ 13. $8\dfrac{1}{3} \div 1\dfrac{2}{3}$

4. $\dfrac{8}{11} \div \dfrac{1}{2}$ 9. $\dfrac{9}{10} \div \dfrac{27}{50}$ 14. $5\dfrac{5}{9} \div 8\dfrac{1}{3}$

5. $\dfrac{2}{3} \div \dfrac{3}{8}$ 10. $2\dfrac{2}{5} \div \dfrac{3}{8}$ 15. $3\dfrac{1}{3} \div 2\dfrac{1}{2}$

ADDITION OF RATIONAL NUMBERS

By using the properties of operations and what we have learned about multiplication, the addition of rational numbers is easy when the denominators are the same.

$$\frac{5}{9} + \frac{2}{9} = \left(5 \times \frac{1}{9}\right) + \left(2 \times \frac{1}{9}\right)$$

$$= (5 + 2) \times \frac{1}{9}$$

$$= 7 \times \frac{1}{9}$$

$$= \frac{7}{9}$$

A comparison of $\frac{5}{9} + \frac{2}{9}$ and $\frac{7}{9}$ reveals that we might have added the numerators and used the same denominator.

$$\frac{5}{9} + \frac{2}{9} = \frac{5 + 2}{9} = \frac{7}{9}$$

In general, for rational numbers $\frac{a}{n}$ and $\frac{c}{n}$,

$$\frac{a}{n} + \frac{c}{n} = \frac{a + c}{n}.$$

Furthermore, knowing that addition is associative, we can state the following.

For all rational numbers $\dfrac{a}{n}$, $\dfrac{c}{n}$, and $\dfrac{e}{n}$,

$$\frac{a}{n}+\frac{c}{n}+\frac{e}{n}=\frac{a+c+e}{n}.$$

Exercises 6–9:

Find each sum. State the answer in lowest terms.

1. $\dfrac{2}{3}+\dfrac{1}{3}$ 5. $\dfrac{4}{15}+\dfrac{2}{15}+\dfrac{4}{15}$ 9. $2\dfrac{1}{3}+\dfrac{2}{3}$

2. $\dfrac{5}{11}+\dfrac{6}{11}$ 6. $\dfrac{3}{20}+\dfrac{7}{20}+\dfrac{5}{20}$ 10. $7\dfrac{1}{3}+\dfrac{1}{3}$

3. $\dfrac{3}{8}+\dfrac{5}{8}$ 7. $\dfrac{1}{9}+\dfrac{2}{9}+\dfrac{3}{9}$ 11. $1\dfrac{3}{7}+\dfrac{2}{7}$

4. $\dfrac{2}{13}+\dfrac{5}{13}$ 8. $\dfrac{4}{25}+\dfrac{3}{25}+\dfrac{8}{25}$ 12. $3\dfrac{1}{5}+4\dfrac{2}{5}$

Suppose we are to find the sum of $\frac{4}{5}$ and $\frac{2}{3}$. In this case the denominators are not the same. But we know how to rename these numbers so that the denominators are the same. Then we could proceed as in the previous exercises.

$$\frac{4}{5}+\frac{2}{3}=\left(\frac{4}{5}\times 1\right)+\left(\frac{2}{3}\times 1\right)$$

$$=\left(\frac{4}{5}\times\frac{3}{3}\right)+\left(\frac{2}{3}\times\frac{5}{5}\right)$$

$$=\frac{4\times 3}{15}+\frac{5\times 2}{15}$$

$$=\frac{12}{15}+\frac{10}{15}$$

$$=\frac{12+10}{15}$$

$$=\frac{22}{15}$$

But how did we know to choose $\frac{3}{3}$ and $\frac{5}{5}$ as names for 1? Notice that $\frac{4}{5}$ was multiplied by $\frac{3}{3}$ and that 3 is the denominator of the other factor. Is this also true for $\frac{2}{3}\times\frac{5}{5}$? This procedure may not yield the least possible denominator common to both rational numbers.

Definition 6–5:

For all rational numbers $\dfrac{a}{n}$ and $\dfrac{r}{s}$,

$$\frac{a}{n}+\frac{r}{s}=\frac{as+nr}{ns}.$$

For example, let us find the sum of $\frac{7}{8}$ and $\frac{2}{5}$.

$$\frac{7}{8}+\frac{2}{5}=\frac{(7\times 5)+(8\times 2)}{8\times 5}$$

$$=\frac{35+16}{40}$$

$$=\frac{51}{40}$$

Exercises 6–10:

Find each sum. State the answer in lowest terms.

1. $\dfrac{2}{3}+\dfrac{1}{7}$ 5. $\dfrac{2}{9}+\dfrac{3}{7}$ 9. $\dfrac{1}{2}+\dfrac{1}{8}+\dfrac{1}{9}$

2. $\dfrac{2}{9}+\dfrac{1}{3}$ 6. $\dfrac{5}{24}+\dfrac{1}{8}$ 10. $\dfrac{2}{5}+\dfrac{1}{4}+\dfrac{1}{3}$

3. $\dfrac{2}{5}+\dfrac{1}{2}$ 7. $\dfrac{3}{4}+\dfrac{2}{9}$ 11. $\dfrac{1}{12}+\dfrac{1}{5}+\dfrac{3}{8}$

4. $\dfrac{2}{3}+\dfrac{3}{4}$ 8. $\dfrac{3}{11}+\dfrac{2}{7}$ 12. $\dfrac{1}{2}+\dfrac{1}{3}+\dfrac{1}{4}$

MIXED NUMERALS IN ADDITION

It is usually easier to add rational numbers named by mixed numerals in vertical form rather than horizontal form. However, we shall use the horizontal form and the properties we know to verify that the vertical arrangement is valid. Suppose we are to find the sum of 5⅔ and 3½.

$$5\frac{2}{3}+3\frac{1}{2}=\left(5+\frac{2}{3}\right)+\left(3+\frac{1}{2}\right)$$

$$=5+\left(\frac{2}{3}+3\right)+\frac{1}{2}$$

$$=5+\left(3+\frac{2}{3}\right)+\frac{1}{2}$$

$$=(5+3)+\left(\frac{2}{3}+\frac{1}{2}\right)$$

At this point in the solution we notice that we are to find the sum of two whole numbers, the sum of two rational numbers, and then to add these sums. Hence we can state this as follows.

$$5\frac{2}{3} \text{ and think of: } 5+\frac{2}{3}$$
$$+3\frac{1}{2} \qquad\qquad 3+\frac{1}{2}$$
$$\overline{\qquad\qquad (5+3)+\left(\frac{2}{3}+\frac{1}{2}\right)}$$

We already know how to find the required sums in this example.

$$5\frac{2}{3} \qquad\qquad 5\frac{4}{6}$$
$$+3\frac{1}{2} \qquad\qquad +3\frac{3}{6}$$
$$\overline{\qquad\qquad} \qquad\qquad \overline{\qquad}$$
$$\qquad\qquad\qquad 8\frac{7}{6}$$

Since ⅞ = 6/6 + ⅙ = 1 + ⅙, we can restate the sum as follows.

$$8\frac{7}{6}=8+\frac{7}{6}=8+\left(1+\frac{1}{6}\right)=(8+1)+\frac{1}{6}=$$

$$9+\frac{1}{6}=9\frac{1}{6}$$

We say that a mixed numeral is in lowest terms if the fractional part names a number less than one, and if the fraction is in lowest terms.

Exercises 6–11:
Find each sum. State the answer in lowest terms.

1. $6\frac{3}{8}$ $+5\frac{1}{8}$	4. $28\frac{9}{10}$ $+43\frac{4}{5}$	7. $215\frac{3}{4}$ $+82\frac{1}{6}$
2. $18\frac{3}{4}$ $+7\frac{1}{2}$	5. $27\frac{3}{5}$ $+6\frac{5}{8}$	8. $752\frac{3}{7}$ $+54\frac{1}{2}$
3. $12\frac{2}{3}$ $5\frac{1}{4}$ $+3\frac{1}{2}$	6. $9\frac{7}{8}$ $13\frac{1}{4}$ $+52\frac{1}{2}$	9. $115\frac{1}{6}$ $607\frac{2}{5}$ $+79\frac{1}{4}$

SUBTRACTION OF RATIONAL NUMBERS

Most of this chapter thus far has used primarily the positive integers and hence positive rational numbers since they receive the greater attention in arithmetic. However, the definitions for multiplication, division, and addition of rational numbers apply as well to the negative rational numbers.

In subtraction of integers we defined subtraction as follows: For all integers a and b, $a-b$ means $a+(-b)$. That is, to subtract a number, we added its opposite or additive inverse. This same idea applies to subtraction of rational numbers. For example:

$$\frac{5}{8}-\frac{3}{8}=\frac{5}{8}+\left(-\frac{3}{8}\right).$$

Before completing this example, let us investigate how we might rename −⅜. By the property of −1, −⅜ = (−1) × ⅜. But we already know that -1 can be named by $-\frac{1}{1}$ or $\frac{1}{-1}$. Hence, we can use either of these in place of -1, as shown below.

$$-\frac{3}{8} = (-1) \times \frac{3}{8} \qquad -\frac{3}{8} = (-1) \times \frac{3}{8}$$

$$= \frac{-1}{1} \times \frac{3}{8} \qquad\qquad = \frac{1}{-1} \times \frac{3}{8}$$

$$= \frac{(-1) \times 3}{1 \times 8} \qquad\qquad = \frac{1 \times 3}{(-1) \times 8}$$

$$= \frac{-3}{8} \qquad\qquad\qquad = \frac{3}{8-}$$

Hence, $\dfrac{3}{8} = \dfrac{-3}{8} = \dfrac{3}{-8}$.

In fact, for any rational number $-\dfrac{a}{n}$,

$$-\frac{a}{n} = \frac{-a}{n} = \frac{a}{-n}.$$

By using this idea we can always rename a rational number so the sign preceding the fraction is $+$, which we have agreed will be implied when no sign is written.

Now let us return to $\frac{5}{8} - \frac{3}{8}$.

$$\frac{5}{8} - \frac{3}{8} = \frac{5}{8} + \left(-\frac{3}{8}\right)$$

$$= \frac{5}{8} + \frac{-3}{8}$$

$$= \frac{5 + (-3)}{8}$$

$$= \frac{2}{8} \text{ or } \frac{1}{4}$$

We can define subtraction of rational numbers by applying the above idea to the definition of addition of rational numbers.

Definition 6–6:

For all rational numbers $\dfrac{a}{n}$ and $\dfrac{r}{s}$,

$$\frac{a}{n} - \frac{r}{s} = \frac{as - nr}{ns}.$$

Study how this definition is used to find the following difference.

$$3\frac{4}{5} - \frac{2}{3} = \frac{19}{5} - \frac{2}{3} \qquad\qquad 3\frac{12}{15}$$

$$= \frac{(19 \times 3) - (5 \times 2)}{5 \times 3} \qquad \frac{10}{15}$$

$$\qquad\qquad\qquad\qquad\qquad 3\frac{2}{15}$$

$$= \frac{57 - 10}{15}$$

$$= \frac{47}{15} \text{ or } 3\frac{2}{15}$$

Exercises 6–12:

Find each difference. State the answer in lowest terms.

1. $\dfrac{7}{9} - \dfrac{3}{9}$ 6. $\dfrac{3}{5} - \dfrac{2}{7}$ 11. $7\dfrac{1}{6} - 4\dfrac{3}{4}$

2. $\dfrac{7}{15} - \dfrac{2}{15}$ 7. $\dfrac{7}{9} - \dfrac{2}{3}$ 12. $8\dfrac{2}{5} - 6\dfrac{3}{8}$

3. $5 - \dfrac{3}{5}$ 8. $\dfrac{12}{25} - \dfrac{2}{5}$ 13. $45\dfrac{5}{9} - 37\dfrac{5}{6}$

4. $3\dfrac{7}{10} - \dfrac{3}{5}$ 9. $\dfrac{7}{12} - \dfrac{5}{16}$ 14. $86\dfrac{4}{7} - 55\dfrac{7}{8}$

5. $\dfrac{8}{11} - \dfrac{2}{3}$ 10. $\dfrac{7}{8} - \dfrac{4}{5}$ 15. $47\dfrac{4}{9} - 36\dfrac{3}{5}$

DECIMALS

Whenever the denominator is a power of ten or whenever the rational number can be renamed so the denominator is a power of ten, the rational number can easily be named as a *decimal numeral* or simply a *decimal*.

Study how each of the following rational numbers is named as a decimal.

$$\frac{27}{100} = .27$$

$$\frac{3}{25} = \frac{4}{4} \times \frac{3}{25} = \frac{12}{100} = .12$$

$$3\frac{2}{5} = \frac{17}{5} = \frac{2}{2} \times \frac{17}{5} = \frac{34}{10} = 3.4$$

$$\frac{3}{250} = \frac{4}{4} \times \frac{3}{250} = \frac{12}{1000} = .012$$

The digits to the right of the decimal point (.) name the numerator and the number of such digits indicates the power of ten which is the denominator. For example, .347 denotes a numerator of 347 and a denominator of 10^3 or 1000.

Study how the following decimals are rewritten as fractions.

$$.57 = \frac{57}{100}$$

$$.32 = \frac{32}{100} = \frac{4 \times 8}{4 \times 25} = \frac{8}{25}$$

$$5.13 = 5 + .13 = 5 + \frac{13}{100} = 5\frac{13}{100}$$

Exercises 6–13:

Write each of the following as a decimal.

1. $\dfrac{47}{100}$　　3. $\dfrac{17}{25}$　　5. $\dfrac{28}{250}$

2. $\dfrac{4}{5}$　　4. $\dfrac{3}{4}$　　6. $\dfrac{13}{20}$

Write each of the following as a fraction or a mixed numeral.

7. 12.7　　9. .125　　11. .034

8. .0037　　10. .6　　12. .0475

DECIMAL NUMERATION

Base-ten numeration is also called decimal numeration. The scheme of place value is shown in the following diagram.

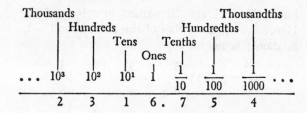

The numeral shown above is read *two thousand three hundred sixteen and seven hundred fifty-four thousandths*. Notice that the digits to the left of the decimal point are read just as for a whole number; the decimal point is read *and*; and the digits to the right of the decimal point are read just as for a whole number and followed by the place-value name of the rightmost digit.

DECIMALS IN ADDITION AND SUBTRACTION

Before discussing how to compute with decimals, it is important to realize that we are still performing operations on rational numbers. The properties, such as commutative and associative properties, are properties of the operations. Hence, we have nothing new to learn about these operations; we only learn the rules for using the decimal numerals when the operations are to be performed.

By using expanded notation for some decimals, the usual pattern for writing decimals for addition or subtraction becomes evident. Suppose we are to find the sum of 34.12 and 47.35.

To make the above addition easy we might write these numerals another way and add those numbers that have the same place value.

$$34.12 = 30 + 4 + \frac{1}{10} + \frac{2}{100}$$

$$+ 47.35 = 40 + 7 + \frac{3}{10} + \frac{5}{100}$$

$$70 + 11 + \frac{4}{10} + \frac{7}{100} = 81.47$$

In this scheme we notice that the numerals for the tens (30 and 40) are aligned vertically, the numerals for the ones (4 and 7) are aligned vertically, and the same is true for the tenths and hundredths.

When writing decimals for addition or subtraction, such an alignment becomes nothing more than aligning the decimal points. Then we add or subtract just as if the numerals named whole numbers. Then write the decimal point in the numeral for the sum or difference directly below those used to name the numbers in the addition or subtraction.

Addition Examples:

34.12	5.907	312.5907
+ 47.35	+ .218	+ 36.4725
81.47	6.125	349.0632

Subtraction Examples:

36.73	7.459	472.3652
− 19.52	− .536	− 83.0764
17.21	6.923	389.2888

Exercises 6–14:

Find each sum.

1.	36.49	3.	248.6	5.	7.26105
	+ 18.72		+ 35.7		+ 5.72964

2.	3.595	4.	.4321	6.	.501247
	.608		.7006		.000986
	+ 7.172		+ .3895		+ .720058

Find each difference.

7.	63.35	9.	80.05	11.	4.7092
	− 19.75		− 26.16		− .8135

8.	926.3	10.	.308	12.	.51246
	− 756.4		− .269		− .36189

DECIMALS IN MULTIPLICATION

Our definition of multiplication of rational numbers stated that we find the product of the numerators and the product of the denominators.

Since decimals are shorthand names for rational numbers, we should expect the same definition to apply.

$$.3 \times .23 = \frac{3}{10} \times \frac{23}{100} = \frac{69}{1000} = .069$$

Since .3 denotes the numerator 3 and the denominator 10, and since .23 denotes the numerator 23 and the denominator 100, we can find the product of the numerators as we do with fractions, and then place the decimal point so that the indicated denominator is the product of 10 and 100. We can state the above multiplication as follows.

$$\frac{\begin{array}{r}.23\\ \times .3\end{array}}{.069}$$

To find the product of 3.201 and .7, think: The numerator will be 3201×7 and the decimal point should be placed so that it indicates a denominator $10^3 \times 10$ or 1000×10 or 10,000. In other words, there should be 4 digits to the right of the decimal point.

$$\frac{\begin{array}{r}3.201\\ .7\end{array}}{2.2407}$$

How many digits are to the right of the decimal point in 3.201? In .7? In 2.2407? Do you see a relationship between these? The number of digits to the right of the decimal point in the numeral for the product is the sum of the numbers of digits to the right of the decimal points in the two factors.

Exercises 6–15:

Find each product.

1.	3.4	3.	84.3	5.	76.29
	× .8		× 2.4		× .34

2.	.76	4.	5.12	6.	315.6
	× 5		× .37		× .215

DECIMALS IN DIVISION

Suppose you are to find the quotient when .35 is divided by .7. Since $\frac{.35}{.7} = \frac{.35}{.7} \times 1$, we can rename 1 so that the product has a denominator that is a whole number. In this case we name 1 as $\frac{10}{10}$ since $.7 \times 10 = 7$ which is a whole number.

$$\frac{.35}{.7} = \frac{.35}{.7} \times 1 = \frac{.35}{.7} \times \frac{10}{10} = \frac{3.5}{7}$$

Since this can be done for any rational number, we need only concern ourselves with the case where the divisor is a whole number.

$$\frac{3.5}{7} = 3.5 \div 7 = \frac{35}{10} \div 7 = \frac{35}{10} \times \frac{1}{7}$$

$$= \frac{35 \times 1}{10 \times 7} = \frac{35 \times 1}{7 \times 10} = \frac{35}{7} \times \frac{1}{10}$$

The final product above shows that we can divide as if both numerator and denominator were whole numbers if we multiply the quotient by $\frac{1}{10}, \frac{1}{100}, \frac{1}{1000}$, and so on, as appropriate.

Another example will help clarify this idea. Suppose we divide .875 by .25. In this case we think as follows.

$$\frac{.875}{.25} = \frac{.875}{.25} \times \frac{100}{100} = \frac{87.5}{25}$$

This step is equivalent to multiplying both the dividend and the divisor by 100. In using the division algorism this would appear as:

$$.25\overline{)\,.875\,} \quad \text{becomes} \quad 25\overline{)\,87.5\,}$$

Then we divide as if both divisor and dividend were whole numbers.

$$\begin{array}{r} 3\,5 \\ 25\overline{)\,87.5\,} \\ 75 \\ \hline 12\,5 \\ 12\,5 \\ \hline \end{array}$$

But how do we know whether to multiply this quotient by $\frac{1}{10}$, or $\frac{1}{100}$, and so on? If 87.5 were written as a fraction, what is the denominator? How many decimal places are to the right of the

decimal point in 87.5? Is there a relationship between these? Hence, we must multiply by $\frac{1}{10}$. Hence the quotient is $35 \times \frac{1}{10}$ or 3.5.

$$\begin{array}{r} 3.5 \\ 25\overline{)\,87.5\,} \end{array}$$

Perhaps you can discover a shortcut for placing the decimal point in the numeral for the quotient.

Exercises 6–16:
Find each quotient.

1. $86.95 \div 4.7$ 3. $6.7332 \div .124$
2. $1.008 \div 3.6$ 4. $3.087 \div .21$

TERMINATING DECIMALS

Now that we have used decimals to name numbers, some interesting questions arise. Can every rational number be named by a decimal? Does every decimal name a rational number?

It is easy to change from a fraction to a decimal since the fraction notation is also considered to denote division.

The following show how to change the fractions $\frac{1}{8}$, $\frac{17}{20}$, and $\frac{4}{25}$ to decimals.

$$\begin{array}{r} .125 \\ 8\overline{)\,1.000\,} \\ 8 \\ \hline 20 \\ 16 \\ \hline 40 \\ 40 \\ \hline 0 \end{array} \qquad \begin{array}{r} .85 \\ 20\overline{)\,17.00\,} \\ 16\,0 \\ \hline 1\,00 \\ 1\,00 \\ \hline 0 \end{array} \qquad \begin{array}{r} .16 \\ 25\overline{)\,4.00\,} \\ 2\,5 \\ \hline 1\,50 \\ 1\,50 \\ \hline 0 \end{array}$$

We say that each of these quotients is named by a *terminating decimal* since we reach a remainder of zero in each case. Notice however, that each divisor (8,20,25) is a factor of 100 or 1000 and hence the rational numbers $\frac{1}{8}$, $\frac{17}{20}$, and $\frac{4}{25}$ could be multiplied by the identity number of multiplication in the form $\frac{n}{n}$ so that the denominator is either 100 or 1000. Therefore, we have been discussing only special kinds of fractions.

Recall that a decimal indicates a denominator

that is a power of 10. Since the least factors of 10 are 2 and 5 (not including 1 and 10 itself), any rational number whose denominator has factors of only 2's and 5's, or both 2's and 5's, can be named as a terminating decimal. If we have a rational number such as

$$\frac{9}{40} = \frac{9}{2 \times 2 \times 2 \times 5}$$

we can make its denominator a power of 10 by pairing each 2 with a 5 (as long as they last) and then supplying any 2's or 5's needed to complete the pairings. The needed 2's and 5's are supplied by multiplying both numerator and denominator by the same number.

$$\frac{9}{40} = \frac{9}{2 \times 2 \times 2 \times 5} \times \frac{5 \times 5}{5 \times 5}$$

$$= \frac{9 \times 5 \times 5}{2 \times 2 \times 2 \times 5 \times 5 \times 5}$$

$$= \frac{9 \times 25}{(2 \times 5)(2 \times 5)(2 \times 5)}$$

$$= \frac{225}{10 \times 10 \times 10} \text{ or } \frac{225}{1000} \text{ or } .225$$

Hence, we need only inspect the denominator to determine whether or not a rational number can be named by a terminating decimal. For example, $\frac{7}{16}$ can be named by a terminating decimal since $16 = 2 \times 2 \times 2 \times 2$.

Exercises 6–17:

Use division to name each of these as a terminating decimal.

1. $\dfrac{7}{20}$ 3. $\dfrac{11}{40}$ 5. $2\dfrac{3}{4}$

2. $\dfrac{7}{8}$ 4. $\dfrac{19}{50}$ 6. $58\dfrac{9}{25}$

REPEATING DECIMALS

Now let us consider a rational number, such as $\frac{4}{11}$, where the denominator has factors other than 2 or 5.

$$\begin{array}{r} .36 \\ 11\overline{)4.0000} \\ \underline{33} \\ 70 \\ \underline{66} \\ 4 \end{array}$$

Can you tell without further division the next two digits of the quotient? Does the decimal naming the quotient ever terminate? A decimal such as .3636 . . . is called a *repeating decimal* or a *periodic decimal* since the digits 36 continue to repeat. In order to name such a decimal more concisely, let us draw a bar over the sequence or period of digits that repeats.

$$\frac{4}{11} = .36\overline{36} \text{ or } .\overline{36}$$

$$\frac{1}{3} = .33\overline{3} \text{ or } .3\overline{3} \text{ or } .\overline{3}$$

The number of times we repeat the period of digits before drawing a bar over it is immaterial. One repetition is sufficient.

Think about naming $\frac{5}{7}$ as a decimal. What are the possible remainders when the divisor is 7? (0,1,2,3,4,5,6) This means that after annexing 0's to the dividend numeral, a remainder must repeat in at most 7 division steps. If a remainder of zero occurs, the decimal terminates. As soon as a remainder repeats, we can stop dividing since we know what the repeating decimal is.

Thus, we see that every rational number can be named by either a terminating or a repeating decimal.

Exercises 6–18:

Use division to name each of these as a repeating decimal.

1. $\dfrac{5}{6}$ 3. $\dfrac{2}{3}$ 5. $\dfrac{11}{14}$

2. $\dfrac{1}{9}$ 4. $\dfrac{5}{33}$ 6. $\dfrac{4}{7}$

CHANGING DECIMALS TO FRACTIONS

There is no problem in changing a terminating decimal to a fraction. Merely reading the decimal tells us the fraction. For example, we read .213 as *two hundred thirteen thousandths* and write the fraction $213/1000$. Since this is already in lowest terms, we leave it as it is.

How can we change a repeating decimal to a fraction? For example, let us change .23 to a fraction. Think of .23 as naming some rational number k.

$$k = .\overline{23}$$
$$\text{Then } 100k = 23.\overline{23}$$

If we should subtract k from $100k$, we should eliminate the decimal and need work only with whole numbers.

$$100k = 23.\overline{23}$$
$$k = .\overline{23}$$
$$\overline{99k = 23}$$
$$k = \frac{23}{99}$$

Since $23/99$ is already in lowest terms we conclued that $.23 = 23/99$.

Another example might be to name .618 as a fraction. Since the digits repeat in a period of 3 digits, we will multiply by 1000 in this case.

$$\text{Let } t = .\overline{618}$$
$$\text{Then } 1000t = 618.\overline{618}$$
$$t = .\overline{618}$$
$$\overline{999t = 618}$$
$$t = \frac{618}{999} \text{ or } \frac{206}{333}$$

We conclude that $.618 = 206/333$.

Thus, we see that every terminating or repeating decimal can be changed to a fraction.

Can you think of a decimal that never repeats nor terminates? How about .010110111 . . . or .343343334 . . . ? These decimals name numbers that are *not* rational. Such numbers are called *irrational numbers*. You have already worked with some irrational numbers, such as $\sqrt{2}$, $\sqrt{5}$, and π. The union of the set of rational numbers and the set of irrational numbers is the set of *real numbers*.

Exercises 6–19:
Change each of these decimals to a fraction.

1. $.35\overline{35}$ 3. $.\overline{396}$ 5. $.7\overline{33}$
2. $5.6\overline{6}$ 4. $.\overline{6315}$ 6. $2.26\overline{12}$

DENSITY OF RATIONAL NUMBERS

Certainly there are no integers between 2 and 3 or between 0 and −1. We can summarize this by saying that there is no integer between two consecutive integers.

We might ask: Is there a rational number between any two given rational numbers? For example, is there a rational number between ¼ and ¾? In this case we can easily see that $2/4$ or ½ is between these two rational numbers.

Then we might ask if there is a rational number between ¼ and ½. Since $¼ = 2/8$ and $½ = 4/8$ this question is the same as asking if there is a rational number between $2/8$ and $4/8$. Now it is more obvious that $2/8 < 3/8 < 4/8$ (which is another way of saying that $3/8$ is between $2/8$ and $4/8$).

We could continue this line of questioning and become quite convinced that between any two rational numbers there is always a third rational number. But let us investigate a general case on a number line. Let a and c represent any two rational numbers such as $a < c$.

Now let us locate the point for $a + c$.

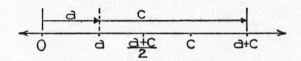

Next, let us find the point midway between the points for 0 and $a + c$, as shown above by $\frac{a+c}{2}$. Is this point between the points for a and c? This process could be used for any two rational numbers.

The rational number $\dfrac{a+c}{2}$ is the average of the two given rational numbers a and c. Hence, to find a rational number between two given rational numbers, find the average of the two given numbers. Thus we may conclude:

Between any two rational numbers there is always a third rational number. We say that the rational numbers are *dense,* or this idea is called the *density property* of the rational numbers.

Exercises 6–20:

Answer the following questions Yes or No.

1. Is there a next greater integer than 1?
2. Is there a next greater rational number than 1?
3. Is there a least positive integer?
4. Is there a least positive rational number?
5. Is there a greatest integer?
6. Is there a greatest rational number?

SETS OF POINTS

THE LINES

Our environment contains many objects from which we develop mathematical ideas—especially our ideas about geometry. This has been true through the ages. The early Egyptians thought a geometrical fact had to be discovered by measurement. To them a taut rope was a straight line. The Greeks were more imaginative and felt that there were no objects that represented a straight line. To them a straight line could exist only in the imagination—or in the make-believe world of geometry.

Mathematicians of today regard geometry much as the Greeks did in those days long ago. We use physical objects as models of geometric ideas. The best we can do is to draw pictures of the things we think about. But this is not strange—we never saw or touched a number, yet we look at sets and invent ideas that we call numbers, we write names for them, and we write phrases about operations on them. In much the same way, we look at physical objects and invent ideas that we then call points, lines, and planes.

The edge of a ruler or a tightly stretched wire serve as models of what we call a *line* (meaning straight line). A line is the property that these objects have in common. A line has no width, no thickness, but infinite length. A line is an idea.

The tip of a needle or a particle of dust in the air serve as models of what we call a *point*. A point is also an idea about an exact location. It has no thickness, no width, no length. It has no dimension whatsoever.

We usually draw dots to represent points. We label our drawings or dots with capital letters and refer to them as point A, point B, and so on.

A •

B •   C •

To represent a line we make a drawing as shown below. We label two points on the line in order to refer specifically to that particular line. We refer to the following as line AB.

The arrows at either end of the drawing serve to remind us that a line extends indefinitely in both directions. We symbolize the name for line AB as $\overleftrightarrow{AB}$. We think of a line as an infinite set of points.

LINE SEGMENTS AND RAYS

We may not be concerned with the entire line, but only part of the line. In other words, we may not be concerned with the entire set of points on a line, only with a particular subset of these points.

We may be interested in only the points C and D and all points on $\overleftrightarrow{CD}$ that lie between points C and D.

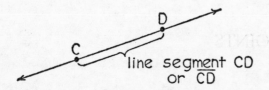

line segment CD
or $\overline{CD}$

A line segment has a definite length—the distance from C to D. Points C and D are called the *end points* of $\overline{CD}$. Certainly line segment CD can also be called line segment DC or $\overline{DC}$. Similarly, $\overleftrightarrow{CD}$ and $\overleftrightarrow{DC}$ are names for the same line.

Another important subset of a line is called a *ray*. In the drawing below, ray CD (denoted by $\overrightarrow{CD}$) consists of point C and all the points on line CD that are on the same side of point C as point D.

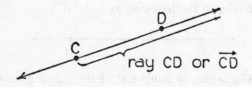

ray CD or $\overrightarrow{CD}$

Point C is called *the* end point of $\overrightarrow{CD}$ since it has only one end point. The arrow indicates that the ray extends indefinitely in this one particular direction.

Of course, we may also think of ray DC (denoted by $\overrightarrow{DC}$).

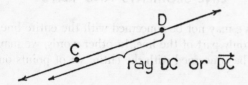

ray DC or $\overrightarrow{DC}$

In this case point D is the end point of ray DC.

Exercises 7–1:

Use the points pictured below to make the following drawings.

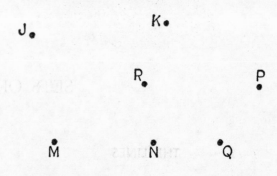

1. Draw line JK.
2. Draw line segment MN.
3. Draw ray RJ.
4. Draw ray PQ.

Answer the following questions Yes or No.

5. Can you measure the length of a line?

6. Can you measure the length of a line segment?

7. Can you measure the length of a ray?

8. Does a line have end points?

9. In the drawing below, do $\overleftrightarrow{AB}$ and $\overleftrightarrow{AC}$ name the same line?

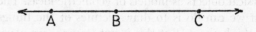

10. In the drawing for Question 9, do $\overrightarrow{AB}$ and $\overrightarrow{BC}$ name the same ray?

ASSUMPTIONS ABOUT POINTS AND LINES

By *space* we mean the set of all points—that is, the set of all exact locations you can possibly think of. Hence, a point or a line is a subset of space. In fact, the study of elementary geometry deals with the relationships that exist between the various subsets of space.

Think about some point A in space. How many lines are there that pass through point A? Use the following drawing as a model.

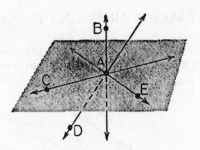

Are there still more lines through point A? How many?

Assumption 1:

There are an unlimited number of lines through any point.

Now think of any two points in space. How many lines pass through both of these points? What happens when you attempt to draw more than one line through both of the points?

Assumption 2:

There is exactly one line through any two given points, or two points determine one and only one line.

Now consider the case where two lines intersect. Our geometrical concept of intersection is the same as that for sets since a line is a set of points. The following is a drawing of two lines that intersect.

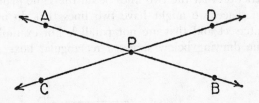

How many points are common to both $\overleftrightarrow{AB}$ and $\overleftrightarrow{CD}$? Hence, the intersection of $\overleftrightarrow{AB}$ and $\overleftrightarrow{CD}$ is point P, which means that the intersection is the set whose only member is point P.

Assumption 3:

Two lines can intersect in at most one point.

Let us use the following drawing to investigate the intersections of some of the subsets of a line.

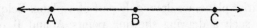

What is the intersection of $\overrightarrow{AC}$ and $\overline{AC}$? In other words, what points do $\overrightarrow{AC}$ and $\overline{AC}$ have in common?

To answer this question, think about the drawing below.

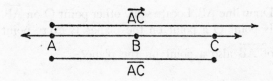

We see that $\overline{AC}$ contains all the points common to both $\overrightarrow{AC}$ and $\overline{AC}$. Hence, $\overrightarrow{AC} \cap \overline{AC} = \overline{AC}$.

Exercises 7–2:

Use the following drawing to complete the sentences below.

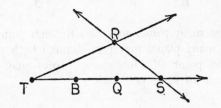

1. $\overline{BQ} \cap \overline{QS} =$ ___
2. $\overleftrightarrow{RS} \cap \overline{RS} =$ ___
3. $\overline{TB} \cap \overline{BQ} =$ ___
4. $\overline{TB} \cap \overline{QS} =$ ___
5. $\overline{BQ} \cup \overline{TB} =$ ___
6. $\overrightarrow{QS} \cup \overrightarrow{QS} =$ ___
7. $\overleftrightarrow{RS} \cap \overrightarrow{TS} =$ ___
8. $\overrightarrow{TS} \cap \overrightarrow{TR} =$ ___

PLANES

The surface of a windowpane or the surface of a table top or any other flat surface serves as a model of what we call a *plane*. A plane has no thickness, but extends indefinitely without boundary. Obviously a plane consists of many exact locations (points) and is a subset of space.

Since it is impossible to draw a picture of an entire plane, we draw only part of a plane. Below is such a drawing showing points A and B on the plane.

Draw line AB. Locate some other point Q on $\overleftrightarrow{AB}$. Is point Q a point on the plane? Is every point of $\overleftrightarrow{AB}$ also a point on the plane?

Assumption 4:

If two points of a line are on a plane, then the line is on the plane.

Think of three points that are not on the same line, such as

C•

A•   •B

How many planes may pass through point A? How many planes may pass through both point A and point B? How many planes may pass through all three points?

Assumption 5:

Three points that are not on the same line determine one and only one plane.

Think of the plane of one wall of a room and the plane of the ceiling. Do they intersect? What kind of geometric figure is their intersection?

Assumption 6:

Two planes can intersect in at most one line.

PARALLEL LINES AND PLANES

The following drawing shows two lines on a plane which do not intersect, or their intersection is the empty set.

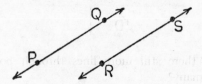

We say that $\overleftrightarrow{PQ}$ is parallel to $\overleftrightarrow{RS}$ and denote this by $\overleftrightarrow{PQ} \parallel \overleftrightarrow{RS}$.

If two lines are on the same plane one and only one of the following situations must exist.

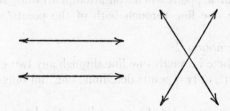

Assumption 7:

Two lines in the same plane are either parallel lines or intersecting lines.

Notice that the above discussions have the restriction that the two lines be on the same plane. In space, we might have two lines that do not intersect and that are not parallel. For example, the drawing below shows a rectangular box.

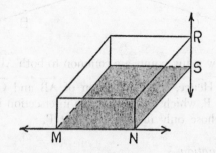

Lines MN and RS would never intersect, but we would not call them parallel lines. They are called *skew lines.*

Think about the planes of the floor and the ceiling of a room. Do these planes intersect?

Definition 7–1:

Two lines in a plane, two planes, or a line and a plane are *parallel* if their intersection is the empty set.

Exercises 7–3:

Letter *m* and *n* refer to the two planes shown in the following figure. For each item 1 through 10 below, state the letter of the corresponding item in *a* through *j*.

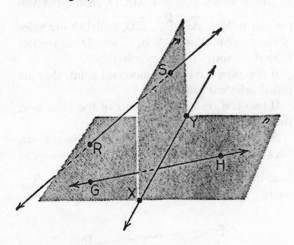

1. Two skew lines

2. Intersection of $\overline{RS}$ and $\overleftrightarrow{RS}$
3. Intersection of planes *m* and *n*

4. Intersection of $\overrightarrow{SR}$ and $\overleftrightarrow{XY}$
5. Line segment on plane *n*
6. Two intersecting lines

7. Intersection of plane *n* and $\overleftrightarrow{RS}$
8. A point not on plane *n*

9. Intersection of $\overrightarrow{GH}$ and $\overrightarrow{HG}$
10. A point common to planes *m* and *n*

a. GH   d. $\overleftrightarrow{XY}$ and $\overleftrightarrow{GH}$ g. $\overline{RS}$

b. $\overline{XY}$ or $\overline{GH}$ e. Ø   h. point R

c. $\overleftrightarrow{XY}$ f. $\overleftrightarrow{RS}$ and $\overleftrightarrow{XY}$ i. point X

                                         j. point S

SEPARATION PROPERTIES

Another important relationship among points, lines, and planes is what we call *separation*. Think of the plane represented by a wall of a room. This plane separates space into three distinct sets of points—those points on one side of the plane, those on the other side of the plane, and those on the plane. Those points on one side of the plane are usually referred to as a *half-space*. The plane itself is not in either half-space.

The following figure shows plane *m* separating space into three distinct sets of points: the half-space containing point R, the half-space containing points S and T, and the plane itself.

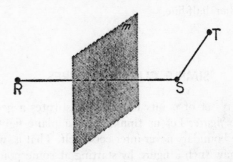

Points S and T are in the same half-space since they can be joined by a line segment that does not intersect plane *m*. Since line segment RS intersects plane *m* we say that points R and S are in different half-spaces.

In the same way, a line separates a plane into three distinct sets of points—the line and the two *half-planes*. In the following figure line AB separates the plane *n* into the half-plane containing points G and H, the half-plane containing point K, and the line AB.

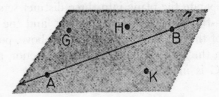

Since $\overline{GH} \cap \overleftrightarrow{AB} = \emptyset$, points G and H are in the same half-plane. Since $\overline{GK}$ would intersect $\overleftrightarrow{AB}$, we say that points G and K are in different half-planes.

A similar situation exists for a point and a line.

In the following figure point P separates $\overleftrightarrow{AB}$ into two *half-lines* and the point P.

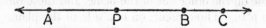

Since $\overline{BC}$ does not contain P, we say that points B and C are on the same half-line. Since $\overline{AB}$ contains P, we say that points A and B are on different half-lines. Point P does not belong to either half-line.

SIMPLE CLOSED FIGURES

Every set of points in space constitutes a *geometric figure*. Let us think about a plane figure whose boundary never intersects itself. That is, we can draw such a figure by starting at some point in a plane, never lift the pencil tip from the paper, and trace any kind of curve we desire, and end at the starting point. Examples of such figures are shown below.

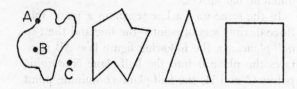

These are called *simple closed figures* because they separate the plane into three distinct sets of points—the figure itself, the interior, and the exterior of the figure. In the first figure above, point A is on the figure, point B is in the interior, and point C is in the exterior.

Definition 7–2:

A *polygon* is a simple closed figure formed by the union of line segments that have common end points. Each line segment is called a *side* of the polygon. The common end point of two sides is called a *vertex* of the polygon.

Study how this definition is applied to the following polygon.

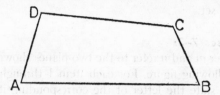

This is called polygon ABCD (name the vertices in order). $\overline{AB}$, $\overline{BC}$, $\overline{CD}$, and $\overline{DA}$ are sides of the polygon. Points A, B, C, and D are vertices (plural of vertex) of the polygon.

If two sides have a common end point, they are called *adjacent sides*.

If two vertices are end points of the same side, they are called *adjacent vertices*.

A line segment joining two nonadjacent vertices of a polygon is called a *diagonal* of the polygon. In the following figure $\overline{SQ}$ and $\overline{PR}$ are diagonals.

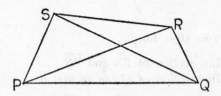

We can now define the various kinds of polygons according to the number of sides each contains.

Definition 7–3:

Kind of Polygon	Number of Sides
Triangle	3
Quadrilateral	4
Pentagon	5
Hexagon	6
Heptagon	7
Octagon	8

Answer the following.

1. A polygon has 6 sides. How many vertices does it have?

2. Is the numeral 8 a simple closed figure? Why or why not?

3. Which kind of polygon has no diagonals?

4. How many diagonals does a quadrilateral have?

5. How many diagonals does a pentagon have?

CIRCLES

Think of locating points in a plane that are a given distance from some point in the plane, as shown in the figure at the left below.

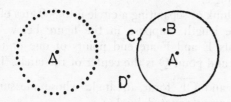

We might continue to locate more and more points the given distance from A. If we think of locating all such points we would be thinking about a circle.

Definition 7–4:

A *circle* is the set of all points in a plane at a given distance from some point in the plane.

In the figure above, point A is called the *center* of the circle; it is not a point of the circle. Point C is on the circle; point B is in the interior; and point D is in the exterior.

Definition 7–5:

Any part of a circle containing more than one point is called an *arc* of the circle.

In the following circle, arc AB (denoted $\overset{\frown}{AB}$) is the set of all points on the circle between and including points A and B.

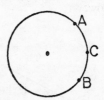

Of course, the above description does not tell which arc is meant. Therefore, let us agree to mean the shorter of the two arcs. Or we might label another point C on the arc and refer to $\overset{\frown}{AB}$ as $\overset{\frown}{ACB}$.

ANGLES

Another important geometric figure or set of points is called an angle.

Definition 7–6:

An *angle* is the union of two rays that have the same end point.

The following are pictures of angles.

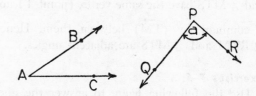

The first figure is called angle BAC (denoted by ∠ BAC) or angle CAB (denoted by ∠ CAB. Notice that point A is the common end point of the two rays; it is called the *vertex* of the angle. Rays AB and AC are called the *sides* of the angle. In naming an angle we usually name a point on one side, the vertex, and then a point on the other side. If no confusion will arise, we may name an angle by giving only the letter of the vertex. That is, ∠ BAC might be called ∠ A.

Another way of naming an angle is to draw a small curved arrow between the two sides and write a lower-case letter, as shown in ∠ QPR above. This angle can then be called ∠ *a*.

An angle separates a plane into three distinct sets of points—the angle, its interior, its exterior —as shown below.

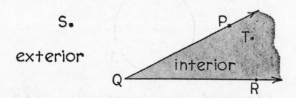

In the following figure, $\overleftrightarrow{RS}$ and $\overleftrightarrow{MN}$ intersect in point T.

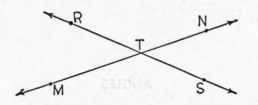

A pair of angles such as ∠ RTM and ∠ NTS are called *vertical angles*. What do these two angles have in common? Are angles RTN and MTS also a pair of vertical angles?

If two angles have the same vertex and a common ray between them, the angles are called *adjacent angles*. In the preceding figure ∠ RTM and ∠ MTS have the same vertex (point T) and a common ray ($\overrightarrow{TM}$) between them. Hence, ∠ RTM and ∠ MTS are adjacent angles.

Exercises 7–5:

Use the following figure to answer the questions below.

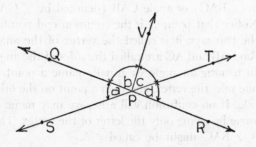

1. Why are ∠ QPT and ∠ VPT not adjacent angles?

2. Name a point in the interior of ∠ QPT.

3. Name a point in the exterior of ∠ QPT.

4. ∠ *a* and ∠ __ are a pair of vertical angles.

5. ∠ *b* and ∠ __ or ∠ __ are a pair of adjacent angles.

6. Is every point of $\overrightarrow{PV}$ in the interior of ∠ QPT? Why or why not?

7. What are the sides of ∠ TPR?

MEASURING ANGLES

The size of an angle depends on the amount of opening between the sides (rays) of the angle. We need a suitable unit for measuring the size of an angle.

Think of separating a circle into 360 arcs of the same length. Suppose in the figure below that points E and F are end points of one of these arcs and point Q is the center of the circle. Then $\overrightarrow{QE}$ and $\overrightarrow{QF}$ form an angle whose measure is 1 *degree* (denoted by 1°).

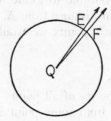

A model showing such a separation for one half of a circle is called a *protractor*. A protractor is an instrument used to measure angles just as a ruler is an instrument used to measure line segments.

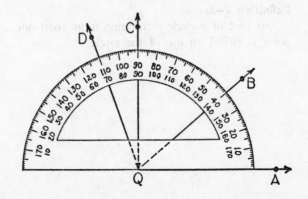

Notice that the center of the circle is marked on the protractor (point Q in the figure). Notice also how the protractor is placed to measure an angle—the center of the circle is placed on the vertex of the angle and the straight edge of the protractor is placed on one side of the angle. We say that the *degree measure* of ∠ AQB is 40. Let us agree to use the symbol m ∠ AQB to mean *the degree measure of ∠ AQB*. Then we can say that m ∠ AQB = 40.

Exercises 7–6:

Use a protractor to find the degree measures of the following angles.

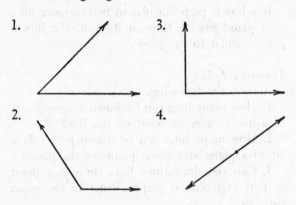

A convenient way of indicating the degree measure of an angle is shown below.

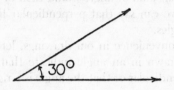

Exercises 7–7:

Use the following figure for the exercises below.

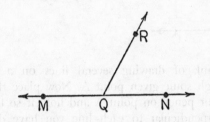

1. What kind of an angle is ∠ MQN? ∠ MQR? ∠ NQR?
2. What is the degree measure of ∠ MQR? Of ∠ NQR? What is the sum of their measures?
3. What is your conclusion about $\overrightarrow{QM} \cup \overrightarrow{QN}$?

KINDS OF ANGLES

Angles are classified according to their degree measure. The angles referred to in the following definition are shown in the previous drawing of a protractor.

Definition 7–7:

A *right angle* has a measure of 90°. ∠ AQC is a right angle.

An *acute angle* has a measure greater than 0° but less than 90°. ∠ AQB is an acute angle.

An *obtuse angle* has a measure greater than 90° but less than 180°. ∠ AQD is an obtuse angle.

A *straight angle* has a measure of 180°.

PERPENDICULAR LINES AND PLANES

If two lines intersect so that all four angles have the same measure we say the lines are *perpendicular.*

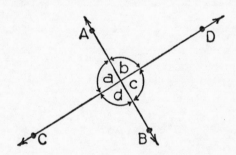

If m ∠ *a* = m ∠ *b* = m ∠ *c* = m ∠ *d*, then $\overleftrightarrow{AB}$ is perpendicular to $\overleftrightarrow{CD}$ (denoted by $\overleftrightarrow{AB} \perp \overleftrightarrow{CD}$) and also $\overleftrightarrow{CD} \perp \overleftrightarrow{AB}$.

On the other hand, if the two lines are perpendicular, then all four angles have the same measure. What kind of angle would each of them be? Hence, we can say that perpendicular lines form right angles.

As a convenience in our drawings, let the symbol ⌐ drawn in an angle indicate that the rays are perpendicular or that the angle is a right angle.

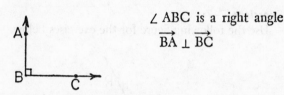

∠ ABC is a right angle

$\overrightarrow{BA} \perp \overrightarrow{BC}$

Think of drawing several lines on a plane through some given point A. Now place the tip of your pencil on point A and hold it so that it is perpendicular to each line you have drawn. Can the pencil be in more than one position so that this is true?

If a line intersects a plane in only one point, let us call this point the *foot* of the line. In the following figure, point P is the foot of $\overleftrightarrow{AB}$, and $\overleftrightarrow{AB}$ is perpendicular to lines CD and EF in plane *n*.

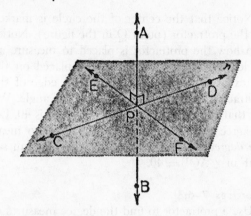

Definition 7–8:

If a line is perpendicular to two or more lines in a plane passing through its foot, the line is perpendicular to the plane.

Exercises 7–8:

Answer the following.

1. How many lines can be drawn perpendicular to a line at a given point on the line?

2. How many lines can be drawn perpendicular to a plane at a given point on the plane?

3. Can you draw three lines through a point so that each line is perpendicular to the other two lines?

4. If $\overleftrightarrow{AB}$ is perpendicular to a vertical line is AB a horizontal line?

5. If $\overleftrightarrow{CD}$ is perpendicular to a horizontal line is $\overleftrightarrow{CD}$ a vertical line?

CHAPTER EIGHT

MEASURE AND MEASUREMENT

WHAT MEASUREMENT IS

Measurement is a connecting link between the physical world and mathematics. We can count the number of buns we need for a picnic, but we measure the amount of milk we need.

To measure a line segment means to assign a number to it. We cannot count all of the points on the line segment since there are infinitely many. So we select some line segment as an arbitrary unit and compare it to the line segment to be measured. Once this concept of measurement is developed, we find it is also useful in determining the measure of other things.

Suppose we want to find the measure of line segment AB shown below. We might choose some other line segment, such as $\overline{MN}$, as the arbitrary unit. By starting at A the unit $\overline{MN}$ can be laid off 4 times to reach point R which is between A and B.

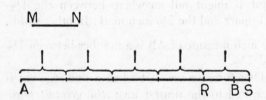

If laid off 5 times we reach point S which is beyond point B. Surely we can say that $\overline{AB}$ has a length *greater than* 4 units and *less than* 5 units. The best we can do is visually estimate that the length of $\overline{AB}$ is nearer to 5 units than to 4 units, and state this by saying that the length of $\overline{AB}$ is 5 units, *to the nearest unit*.

Exercises 8–1:

Use 1 inch as a unit and find the length to the nearest inch of each line segment below.

Then use ½ inch as the unit and find the length of each segment to the nearest ½ inch.

1. _____

2. _____

3. _____

4. _____

APPROXIMATE NATURE OF MEASUREMENT

To help us determine whether the previously used $\overline{AB}$ has a length nearer to 4 units or 5 units, we can bisect the unit $\overline{MN}$.

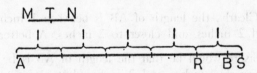

Now we see that point B is more than 4½ units from A, and hence the measurement is nearer to 5 units.

Suppose we use $\overline{MT}$ as the unit of measure. The length of $\overline{AB}$ is 9 units to the nearest unit.

We could continue to bisect each new unit (line segment) and would never find a unit that "fits exactly" a whole number of times into $\overline{AB}$. Similarly, we are not surprised if someone tells us that it is 100 miles to the city of Podunk, and then someone else says it is 98 miles or 103 miles. *Measurement is only approximate, never exact.*

When we say that the "length of a line segment is 5 units" we mean that "the measure of the line segment in terms of some unit is 5." That is, *a measure is a number.* In a phrase such as "26 feet,"

> "26" names the *measure*,
> "feet" names the *unit*, and
> "26 feet" names the *measurement*.

Notice that a measurement contains two symbols—one for the measure and the other for the unit. After all, we have learned how to add, subtract, multiply, and divide only numbers, not units. We do not add or subtract units of measure any more than we add or subtract bananas or horses.

PRECISION

A ruler is simply a model showing different subdivisions of an inch. The figure below shows how we could find the measure of $\overline{AB}$.

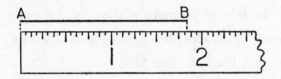

Clearly, the length of $\overline{AB}$ is between 1 inch and 2 inches, and closer to 2 inches. A better measure would be that the length of $\overline{AB}$ is between 1½ inches and 2 inches, but closer to 2 inches. A still better measure would be that the length of $\overline{AB}$ is between 1¾ inches and 2 inches, but closer to 1¾ inches.

If we let x represent the inch measure of AB, we can indicate the value of x as follows.

$$1 < x < 2$$
$$1\frac{1}{2} < x < 2\frac{0}{2}$$
$$1\frac{3}{4} < x < 2\frac{0}{4}$$
$$1\frac{6}{8} < x < 1\frac{7}{8}$$
$$1\frac{13}{16} < x < 1\frac{14}{16}$$

The units on the ruler could be subdivided into still smaller units since between any two points on a line there is a third point.

We say that a measurement of $1\frac{14}{16}$ inches is more *precise* or has greater *precision* than a measure of $1\frac{7}{8}$ inches because a smaller unit of measure is used. Similarly, a measurement of 1¾ inches is more precise than a measurement of 1½ inches.

The smaller the unit of measure, the greater is the precision.

When a measurement is made to the nearest inch, we say the precision of measurement is 1 inch. When a measurement is made to the nearest ½ inch, we say the precision of the measurement is ½ inch.

What does it mean to say that $\overline{AB}$ is 2 inches long, to the nearest inch? It means: If the 0-point of the ruler is placed at point A of AB, then point B might fall anywhere between the 1½-inch mark and the 2½-inch mark. In other words, the inch measure of $\overline{AB}$ is a number between 1½ and 2½.

Therefore, we say that when a line segment is measured to the nearest inch, *the greatest possible error is* ½ inch. This does not mean that we have made a mistake in finding the measure; it simply means that the measurement is approximate within these limits.

We can indicate the greatest possible error of such a measurement by writing $(2 \pm \frac{1}{2})$ inches. The symbol $\pm$ is read *plus or minus*.

The greatest possible error of a measurement is equal to one-half of the unit of measure.

If a measurement is made to the nearest $\frac{2}{4}$ inch, such as $7\frac{2}{4}$ inches, do not reduce the fraction. The fraction $\frac{2}{4}$ indicates that the unit of measure is $\frac{1}{4}$ inch and the greatest possible error is $\frac{1}{8}$ inch. But the fraction $\frac{1}{2}$ indicates that the unit of measure is $\frac{1}{2}$ inch and the greatest possible error is $\frac{1}{4}$ inch. Furthermore, a measurement such as $23\frac{2}{4}$ inches means that the unit of measure is $\frac{1}{4}$ inch while 23 inches means that the unit of measure is 1 inch.

Exercises 8–2:

State the unit of measure, the greatest possible error, and how you would write the measurement to show the greatest possible error for each of the following.

1. $14\frac{1}{2}$ in. 3. $5\frac{7}{8}$ mi. 5. $57\frac{0}{6}$ yd.

2. $3\frac{1}{4}$ ft. 4. 9 in. 6. $15\frac{9}{16}$ in.

DECIMALS DENOTING PRECISION

As the unit of measure becomes smaller and smaller, it is more convenient to use a decimal than a fraction to state the measurement. Since decimal notation is used in virtually all technical, scientific, and business computing, let us consider the greatest possible error and precision as related to decimals.

The precision of a measurement such as 5.27 inches is the place value of the last digit used to name the measure. In this case the precision is .01 inch. The greatest possible error is one half of the unit of measure, and hence one half of the place value of the last digit used to name the measure. In this case, the greatest possible error is $\frac{1}{2} \times .01$ or .005 inch.

If a measurement is stated as 7800 feet it is not clear whether the 0's merely place the decimal point or whether they indicate precision. To avoid confusion of this sort, let us underline the 0 that indicates the precision. For example, 7800 feet means precision of 100 feet; 78_0_0 means precision of 10 feet; and 780_0_ means precision of 1 foot.

When a measurement is stated as 5.320 inches we assume that the unit of measure is .001 inch. Otherwise there would be no purpose in writing the final 0.

Exercises 8–3:

State the unit of measure and the greatest possible error for each of the following.

1. 377.5 ft. 3. .073 in. 5. 5200 ft.

2. 3.85 in. 4. 8.30 ft. 6. 540_0_0 ft.

RELATIVE ERROR AND ACCURACY

A greatest possible error of 1 foot in stating the height of a 100-feet-high building would not be misleading. However, a greatest possible error of 1 foot in stating your height could be misleading. In terms of per cent, the first measurement could be in error by only 1 per cent, while the second measurement could be in error by approximately 16 per cent for a person 6 feet tall.

Hence it is important to know just how inexact a measurement is, or how accurate a measurement is. If we divide the greatest possible error by the measure, the quotient is called the *relative error*.

$$relative\ error = \frac{greatest\ possible\ error}{measure}$$

The smaller the relative error, the greater is the *accuracy* of the measurement. That is, a measurement of 100 feet is more accurate than a measurement of 6 feet.

Study the following examples of finding the relative error of a measurement.

Measure-ment	Greatest Possible Error	Relative Error
25 ft.	.5 ft.	$\frac{.5}{25} = .02 = 2\%$
4.2 in.	.05 in.	$\frac{.05}{4.2} = .0119 = 1.19\%$
.006 in.	.0005 in.	$\frac{.0005}{.006} = .0833 = 8.33\%$

Observe that the last measurement above is the most precise but also the least accurate. Precision depends on the unit of measure, but accuracy depends on the relative error. It is entirely possible for an astronomer to have an error of a million miles in the distance to some galaxy, yet be more accurate than a machinist measuring the diameter of a steel shaft to the nearest .001 inch.

Exercises 8—4:

For each of the following measurements, state (1) the unit of measure, (2) the greatest possible error, and (3) the relative error to the nearest tenth of a per cent.

1. 30 ft.	3. 1.25 mi.	5. 700 mi.
2. 7.5 in.	4. 10.075 in.	6. 2000 yd.

ADDING AND SUBTRACTING MEASURES

Computation involving measures is very important in today's world. Since measurements are always approximate, we should expect the results obtained from computation with measures to also be approximate.

Assume that the length of $\overline{AB}$ below is 3.4 inches and the length of $\overline{BC}$ is 4 inches. How long is $\overline{AC}$?

3.4 in.		4 in.	
A	B		C

Our first notion might be to add these measures $(3.4 + 4 = 7.4)$ and say that $\overline{AC}$ is 7.4 inches long. However, by finding the sums of the minimum values and the maximum values, we can see that this is not the best possible answer.

The measurement 3.4 in. or $(3.4 \pm .05)$ in. indicates an actual length between 3.35 in. and 3.45 in. Similarly, the measurement 4 in. or $(4 \pm .5)$ in. indicates an actual length between 3.5 in. and 4.5 in. Let us find the sums of the minimum values and maximum values.

Minimum Values	Maximum Values
3.35	3.45
+3.5	+4.5
6.85	7.95

Common sense tells us that we cannot improve the precision by computation; it depends entirely on the unit of measure. Hence, our result cannot possibly have precision greater than 1 inch since the measurement 4 in. denotes this precision.

By inspecting the two sums above, we see that 7 in. or $(7 \pm .5)$ in. would seem the most reasonable sum. That is, it would be more representative of the numbers in the interval 6.85 to 7.95 than would 6 in. or 8 in.

There are many rules concerning the accuracy and precision of results obtained from computation with measures. Let us agree to use the following rule when adding or subtracting measures.

The sum or difference of measures cannot be more precise than the least precise measure involved.

Therefore, to add or subtract measures, round the more precise measure so that both measures have the same precision. Then add or subtract the measures just as you would numbers that do not denote measures. Study the following example, where the numbers denote measures.

37.2	Precision is .1
+21.37	Precision is .01

In this case, round 21.37 to the nearest tenth, and add the numbers.

37.2	
+21.4	
58.6	Use this result.

Exercises 8—5:

Assume that the numbers named below denote measures. Find each sum or difference.

1.	51.58	3.	7.007	5.	29000
	+38.8		−3.82		+16000

2.	8.072	4.	9.62	6.	72.0
	+2.4195		−6.001		− 4.78

SIGNIFICANT DIGITS

Which of the following measurements is most accurate: .025 ft., 25 ft., or 250 ft.? To answer the question we could compute the relative error of each measurement. However, the following fractions indicate the necessary divisions and also reveal that we need not do all of the computation to answer the question.

$$\frac{.0005}{.025} \quad \frac{.5}{25} \quad \frac{5}{250}$$

Since multiplying both numerator and denominator of a rational number by the same number does not change the rational number, we see the following relationship.

$$\frac{.0005}{.025} = \frac{.5}{25} = \frac{5}{250}$$

Each of the three measurements .025 ft., 25 ft., and 250 ft. have the same accuracy. In each case only two digits, 2 and 5, are significant. They are called *significant digits*.

Significant digits are directly related to the number of units of measure, as shown in the following table.

Measurement	Unit of Measure	Number of Units	Significant Digits
25 ft.	1 ft.	25	2
.0125 in.	.0001 in.	135	3
23.09 yd.	.01 yd.	2309	4
7200 mi.	100 mi.	72	2

We can state some simple rules for quickly determining the number of significant digits in a numeral.

a. Every nonzero digit is significant.

b. Every 0 between nonzero digits is significant.

c. Every 0 used to denote the precision of measurement is significant.

Let us use these simple rules to determine the number of significant digits in several numerals.

Numeral	Number of Significant Digits	Rule Used
20.27	4	*a, b*
.0036	2	*a*
7500	3	*a, c*
8500	2	*a*
37.720	5	*a, c*
92000	4	*a, b, c*

In .0036 and 8500 the 0's merely place the decimal point and do not denote precision, hence the 0's are not significant.

In 37.720 the 0 denotes precision and is significant.

In 92000 the middle 0 denotes precision and hence the leftmost 0 also denotes precision. Therefore, the two leftmost 0's are significant. You may also think of 92000 as denoting a unit of 10 and hence the number of units is 9200.

Exercises 8–6:

How many significant digits does each of the following numerals have?

1. 47.30	4. .0042	7. 31000
2. 5.436	5. .00425	8. 26000
3. 8.003	6. .0425	9. 47000

MULTIPLYING AND DIVIDING MEASURES

Just as with addition and subtraction of measures, we should like to develop a simple rule to follow when multiplying or dividing measures.

In some cases one factor may be a measure and the other factor an exact number. For example:

A certain kind of pencil is 8.7 in. long. What is the combined length of 4 of these pencils?

An open sentence for the problem is $4 \times 8.7 = \square$. The factor 4 is exact and the factor 8.7 is approximate. If interpreted as repeated addition, we see that the product should be stated to the nearest tenth.

$$4 \times 8.7 = 34.8$$

Hence, the combined length would be given as 34.8 inches.

When both factors are measures we must approach the problem differently. Suppose we are to find the area of a rectangle whose width is 12 inches and whose length is 16 inches. This means that the width is between 11.5 inches and 12.5 inches and the length is between 15.5 inches and 16.5 inches. The following drawing shows the given rectangle, the rectangle of minimum measurements, and the rectangle of maximum measurements.

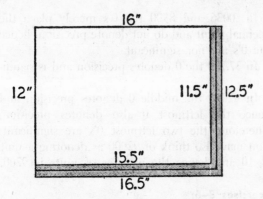

The shaded portion shows the difference between the minimum area and the maximum area. The answer we desire must be somewhere between these two, and preferably stated so that the greatest possible error includes this portion.

We find the area of a rectangle by finding the product of its width and length. Let us compute the area of the three rectangles shown above.

Minimum	Given	Maximum
15.5	16	16.5
×11.5	×12	×12.5
178.25	192	206.25

Obviously, the given area of 192 sq. in. cannot be precise to the nearest square inch since 192 is 13.75 greater than 178.25. It can only be precise to about the nearest 14 square inches.

To overcome indicating the greatest possible error to units 13.75 or 14, we can express 192 sq. in. as 190 sq. in. to the nearest 10 square inches.

There is no satisfactory rule for all cases of multiplying or dividing measures. However, the following is a rough guide for stating the product or the quotient of two measures given by decimal numerals.

The number of significant digits in the product (quotient) of two measures is not greater than the number of significant digits in either factor (in either the dividend or the divisor).

Study how this general rule is used in the following examples.

$$
\begin{array}{ll}
24.3 & \text{(3 significant digits)} \\
\times 2.6 & \text{(2 significant digits)} \\
\hline
145\,8 & \\
486 & \\
\hline
63.18 & \text{(Round to 2 significant digits.)} \\
63 & \text{(Use this answer.)}
\end{array}
$$

$$
\begin{array}{ll}
2.09 & \text{(Round to 2 significant digits.} \\
26\overline{)54.36} & \text{Give the answer as 2.1.)} \\
52 & \\
\hline
2\,36 & \\
2\,34 & \\
\hline
2 &
\end{array}
$$

Exercises 8–7:

Find the product or the quotient of the measures given below.

1. 5.2×43 6. $17.04 \div 24$

2. $.031 \times 2.5$ 7. $.0508 \div 1.2$

3. 7.08×12.3 8. $3700 \div 20$

4. 7500×4.5 9. $502.46 \div 3.15$

5. $.310 \times 12.6$ 10. $2600 \div 42$

THE ENGLISH SYSTEM OF MEASURES

The system of measures most widely used in the United States is called the English system of measures. Some of the common units in this system are inch, foot, yard, mile, ounce, pound, ton, pint, quart, and gallon.

You have probably discovered that the relationships between the various units are not consistent—that is, 12 inches is equivalent to one foot, but 3 feet is equivalent to one yard. Similar irregularities between other units make it difficult to remember all of them. Tables that give these varied relationships can be found in several books. We are forced to be quite familiar with the English system of measures, so let us direct our attention to another very important system of measures.

THE METRIC SYSTEM

In 1789 a group of French mathematicians developed a system of measures that closely resembles the base-ten numeration system. It is so devised that each unit is ten times as large as the next smaller unit. This simplifies the task of converting from one unit to another since it merely requires multiplying or dividing by ten.

The metric system is used almost exclusively in science. It is the most universal system of measures, and is the official system in the United States as well. The extreme cost of changing all of our measuring instruments has delayed its general usage.

Let us center our attention on linear measurement in order to discuss the metric system. The basic unit of linear measure in the metric system is the *meter*.

The French calculated the distance from the Equator to the North Pole along the meridian through Paris. Then they chose $\dfrac{1}{10,000,000}$ of this distance as 1 meter.

Through the great advances in mathematics and science, the meter is today defined in terms of the orange-red wave length of radiating krypton 86 gas.

The common prefixes used in the metric system are given below. They apply to all types of measure.

Prefix	Meaning
micro	$\dfrac{1}{1,000,000}$
milli	$\dfrac{1}{1000}$
centi	$\dfrac{1}{100}$
deci	$\dfrac{1}{10}$
deka	10
hecto	100
kilo	1000
mega	1,000,000

The following table of linear measure in the metric system shows these meanings.

Unit	Abbreviation	Equivalent in Meters
1 millimeter	1 mm.	$\dfrac{1}{1000}$ m.
1 centimeter	1 cm.	$\dfrac{1}{100}$ m.
1 decimeter	1 dm.	$\dfrac{1}{10}$ m.
1 meter	1 m.	1 m.
1 dekameter	1 dkm.	10 m.
1 hectometer	1 hm.	100 m.
1 kilometer	1 km.	1000 m.

How would you convert 7.2 meters to centimeters?

Since 1 m. = 100 cm.
Then 7.2 m. = (100×7.2) cm.
and 7.2 m. = 720 cm.

How would you convert 457.2 mm. to meters?

Since 1 mm. = $\dfrac{1}{1000}$ m.

Then 457.2 mm. = $\left(\dfrac{1}{1000} \times 457.2 \right)$ m.

and 457.2 mm. = .4572 m.

Exercises 8–8:

Convert each of the following measurements as indicated. The equal sign (=) is used here only to denote equivalent measurements, not that the numbers are the same.

1. 6 m. = _____ cm.
2. 834 cm. = _____ m.
3. 5.7 km. = _____ m.
4. 7800 m. = _____ km.
5. 9 m. = _____ mm.
6. 320 mm. = _____ cm.
7. 1 km. = _____ cm.
8. 87.6 cm. = _____ mm.
9. .35 m. = _____ mm.
10. 3580 mm. = _____ m.

COMPARING METRIC AND ENGLISH UNITS

Since both the English and the metric system are used in this country, it is often necessary to **convert** measurements from one system to the other. By comparison, 1 meter is approximately equivalent to 39.37 inches. From this relationship we can obtain other relationships. We shall use the symbol $\approx$ with measurements to mean *is approximately equivalent to.*

$$39.37 \text{ in.} \approx 1 \text{ m.}$$
$$39.37 \text{ in.} \approx 100 \text{ cm.}$$
$$1 \text{ in.} \approx \frac{100}{39.37} \text{ cm.}$$
$$1 \text{ in.} \approx 2.54 \text{ cm.}$$

We can find the approximate English equivalent for 1 centimeter.

$$1 \text{ m.} \approx 39.37 \text{ in.}$$
$$100 \text{ cm.} \approx 39.37 \text{ in.}$$
$$1 \text{ cm.} \approx \frac{39.37}{100} \text{ in.}$$
$$1 \text{ cm.} \approx .39 \text{ in.}$$

Similarly, approximate relationships between 1 yard and 1 meter can be found.

$$1 \text{ yd.} = 36 \text{ in.}$$
$$1 \text{ yd.} \approx \frac{36}{39.37} \text{ in.}$$
$$1 \text{ yd.} \approx .91 \text{ m.}$$

$$1 \text{ m.} \approx 39.37 \text{ in.}$$
$$1 \text{ m.} \approx \frac{39.37}{36} \text{ yd.}$$
$$1 \text{ m.} \approx 1.1 \text{ yd.}$$

How can we convert 3 yards to meters?

$$1 \text{ yd.} \approx .91 \text{ m.}$$
$$(3 \times 1) \text{ yd.} \approx (3 \times .91) \text{ m.}$$
$$3 \text{ yd.} \approx 2.71 \text{ m.}$$

How can we convert 6 inches to centimeters?

$$1 \text{ in.} \approx 2.54 \text{ cm.}$$
$$(6 \times 1) \text{ in.} \approx (6 \times 2.54) \text{ cm.}$$
$$6 \text{ in.} \approx 15.24 \text{ cm.}$$

Exercises 8–9:

Convert each of the following measurements as indicated.

1. 12 in. $\approx$ _____ cm.   5. 1 ft. $\approx$ _____ m.
2. 7 yd. $\approx$ _____ m.   6. 1 ft. $\approx$ _____ cm.
3. 21 cm. $\approx$ _____ in.   7. 23.5 yd. $\approx$ _____ m.
4. 15 m. $\approx$ _____ yd.   8. 17.6 cm. $\approx$ _____ in.

METRIC UNITS OF VOLUME

Just as in the English system, the unit of volume in the metric system is a cube. Furthermore, you may choose whichever cube is convenient for the problem at hand.

If we choose a cube each of whose edges is 1 meter long, the volume is 1 *cubic meter* (1 cu. m.). If we choose a cube each of whose edges is 1 decimeter long, the volume is 1 *cubic decimeter* (1 cu. dm.).

A still smaller, but much used, unit of volume is the cubic centimeter (c.c.). Such a cube is about the size of an ordinary sugar cube. Each edge is 1 cm. or about .4 in. long.

Since we find the volume of a cube by finding the product of its width, height, and length, we can find the relationships between the metric units of volume.

Since 1 m. = 100 cm.,
therefore, 1 cu. m. = $(100 \times 100 \times 100)$ c.c.
or 1 cu. m. = 1,000,000 c.c.

By working the other way we find that 1 c.c. is one-millionth of 1 cu. m.

$$1 \text{ cm.} = \frac{1}{100} \text{ m.}$$

$$\text{so } 1 \text{ c.c.} = \left(\frac{1}{100} \times \frac{1}{100} \times \frac{1}{100}\right) \text{ cu. m.}$$

$$\text{or } 1 \text{ c.c.} = \frac{1}{1,000,000} \text{ cu. m.}$$

Relationships between other metric units of volume can be found in the same way.

$$1 \text{ m.} = 10 \text{ dm.}$$
$$1 \text{ cu. m.} = (10 \times 10 \times 10) \text{ cu. dm.}$$
$$1 \text{ cu. m.} = 1000 \text{ cu. dm.}$$

We can divide both numbers in the last sentence above by 1000 to find that 1 cu. dm. = $\frac{1}{1000}$ cu. m.

Exercises 8–10:

Convert each of the following measurements as indicated.

1. 5 cu. m. = ___ cu. dm.
2. 2730 cu. dm. = ___ cu. m.
3. 8 cu. dm. = ___ c.c.
4. 3850 c.c. = ___ cu. dm.
5. 2.7 cu. m. = ___ c.c.
6. 500,000 c.c. = ___ cu. dm.

CAPACITY

When we speak of the capacity of a container we are usually referring to the volume of the container.

If we say that a particular jar holds 1 gallon of liquid, we are indirectly giving its volume, which incidentally is about 231 cubic inches.

Suppose we have a cube each of whose edges is 1 dm. long.

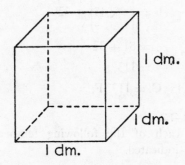

Each edge of this cube would be about 3.9 inches long. The capacity of such a container is very nearly 1 *liter* (pronounced lē'ter), which is the basic unit of capacity in the metric system.

According to today's precise measurements, a liter is equivalent to 1.000028 cu. dm. Unless you measure with precision greater than .0001 liter, you may consider a capacity of 1 liter to be equivalent to 1 cubic decimeter.

Other than the liter, the most often used units of capacity in the metric system are the milliliter and the kiloliter.

1 milliliter (1 ml.) = .0001 liter
1 kiloliter (1 kl.) = 1000 liters

The following statements tell the approximate equivalents between 1 quart and 1 liter.

1 liter (1 l.) ≈ 1.06 quart
1 qt. ≈ .95 l.

To convert from liters to quarts, multiply the number of liters by 1.06.

25 l. ≈ (25 × 1.06) qt.
≈ 26.5 qt.

To convert from quarts to liters, multiply the number of quarts by .95.

14 qt. ≈ (14 × .95) l.
≈ 13.3 l.

Exercises 8–11:
Convert each of the following measurements as indicated.

1. 5 l. = _____ ml.
2. 2800 kl. = _____ l.
3. 49000 ml. = _____ l.
4. 280 l. = _____ kl.
5. 26 qt. ≈ _____ l.
6. 32 l. ≈ _____ qt.

TEMPERATURE

At sea level, the temperatures at which water freezes and boils are taken as fixed points on common thermometer scales. Two common thermometer scales are shown below.

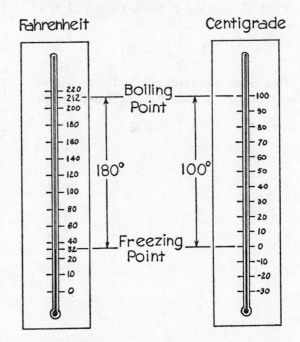

For most daily purposes, such as weather reporting and oven temperatures, the Fahrenheit scale is used. The centigrade scale is used in most scientific work. Many situations arise where it is necessary to convert from one scale to the other.

Notice on the thermometer scales above that 180 Fahrenheit units (called degrees Fahrenheit or °F.) separate the boiling and freezing temperatures of water. On the centigrade scale 100 units (called degrees centigrade or °C.) separate these two temperatures.

We see that 180° F. is equivalent to 100° C., or that for every 180° F. there are 100° C. Furthermore, 1° F. is $\frac{100}{180}$ or $\frac{5}{9}$ as large as 1° C. and

1° C. is $\frac{180}{100}$ or $\frac{9}{5}$ as large as 1° F.

Since 32° F. and 0° C. name the freezing temperature of water, we see that the Fahrenheit scale has a "head start" of 32° over the centigrade scale. When we wish to convert from one scale to the other we must compensate for this head start by subtracting 32 from the Fahrenheit reading or adding 32 to the centigrade reading.

To convert from Fahrenheit to centigrade, we subtract 32 from the F. reading and find ⅝ of the result.

$$C = \frac{5}{9}(F - 32)$$

To convert from centigrade to Fahrenheit, we find ⅞ of the C. reading and add 32 to the result.

$$F = \frac{9}{5}C + 32$$

Let us convert 104° F. to centigrade.

$$C = \frac{5}{9}(104 - 32)$$
$$= \frac{5}{9} \times 72$$
$$= 40$$

Therefore, 104° F. = 40° C.

Now let us convert 45° C. to Fahrenheit.

$$F = \left(\frac{9}{5} \times 45\right) + 32$$
$$= 81 + 32$$
$$= 113$$

Therefore, 45° C. = 113° F.

Exercises 8–12:

Convert each of the following temperature readings as indicated.

1. 200° C. = _____ ° F.   4. 131° F. = _____ ° C.
2. 95° F. = _____ ° C.   5. 115° C. = _____ ° F.
3. 75° C. = _____ ° F.   6. 68° F. = _____ ° C.

CHAPTER NINE

RATIO, PROPORTION, PER CENT

RATIO

We have already discussed the idea of a one-to-one matching or a one-to-one correspondence between two sets. Such a correspondence between sets establishes a relationship between the sets known as equivalence.

Obviously, two sets might have different numbers and hence a different relationship. The following drawing shows a two-to-three correspondence of stars to triangles. That is, for every two stars there are three triangles.

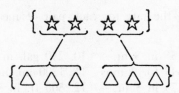

Such a relationship between the numbers of two sets is called a *ratio*. The ratio of the number of stars to the number of triangles is 2 to 3. This ratio can also be denoted as $2:3$ or $\frac{2}{3}$.

The ratio of the number of triangles to the number of stars is 3 to 2 or $3:2$ or $\frac{3}{2}$.

The ratio 2 to 3 simply expresses a relation between the numbers of two sets. It does not state how many objects are in either set. The only essential condition for a ratio of 2 to 3 is that if the first set contains $2k$ objects, then the second set contains $3k$ objects. If $k = 1$, then the first set has 2 objects and the second set contains 3 objects. If $k = 2$, then the first set contains 4 objects and the second set contains 6 objects. If $k = 50$, then the first set contains 100 objects and the second set contains 150 objects.

Notice that a ratio involves a pair of whole numbers, and that their order is important. We have inherited this idea from the early Greeks. They were especially interested in geometry and thought they could compare the sizes of two geometric figures in terms of only whole numbers.

Suppose you were told that the ratio of the number of pounds of bluegrass seed to the number of pounds of crested wheat seed in a certain mixture is 3 to 5. All this tells you is that for every 3 pounds of bluegrass seed there is 5 pounds of crested wheat seed.

If you were further told that a sack of this seed mixture weighed 40 pounds, then you could determine the number of pounds of each kind of seed. That is, $\frac{3}{8}$ of the seed (by weight) is bluegrass seed and $\frac{5}{8}$ of it is crested wheat seed.

Exercises 9–1:

Use the information in statements *a* and *b* to answer the questions which follow:

(*a*) Mr. Smith won the election by a ratio of 3 to 2.

1. If there were 875 votes cast, how many votes did Mr. Smith receive?

2. If Mr. Jones, the only opponent in the election, received 450 votes, how many votes did Mr. Smith receive?

3. If Mr. Smith received 1500 votes, how many votes did his opposition receive?

(*b*) A survey revealed that 5 out of 7 people preferred Brand K over Brand Q.

4. If 490 people were interviewed, how many preferred Brand K?

5. If 600 people preferred Brand K, how many people preferred Brand Q?

DENOTING A RATIO

For computational purposes it is convenient to denote a ratio as a fraction. Since a fraction may be used to denote both a ratio and a rational number, we find that what we have learned about operating with rational numbers is also used to compute with ratios. Similarly, a ratio is usually given as a fraction in simplest terms. For example, we might say that the ratio of the numbers of two sets is 10 to 25. It is usually more meaningful to state this ratio in simple terms.

$$\frac{10}{25} = \frac{5 \times 2}{5 \times 5} = \frac{5}{5} \times \frac{2}{5} = 1 \times \frac{2}{5} = \frac{2}{5}$$

Exercises 9–2:

Express each of the following ratios as a fraction in simplest terms.

1. 8 to 12   5. 12 to 8
2. 81 to 18   6. 175 to 15
3. 35 to 105   7. 144 to 156
4. 16 to 44   8. 30 to 36

RATIO AND MEASUREMENTS

A ratio is defined as a relationship between the *numbers* of two sets. A measurement consists of both a number and a unit of measure. Hence, the comparison of two measurements, such as 4 feet to 9 inches, can be stated as a ratio only if the two measurements are stated with the same unit of measure. In this case we could state both measurements in either feet or inches and state the ratio accordingly.

In feet	In inches
4 to $\frac{3}{4}$	48 to 9
$\frac{4}{\frac{3}{4}} = \frac{16}{3}$	$\frac{48}{9} = \frac{16}{3}$

When we attempt to find the ratio of 20 miles to 4 minutes, we find there is no common unit of measure because one is a linear measurement and the other a time measurement. Of course we could find the ratio of the number 20 to the number 4.

20 mi. to 4 min. or 5 mi. to 1 min.

In the latter of these the number of units is 1 and the units of measure are different. Such a comparison as 5 mi. to 1 min. is called a *rate*. In expressing a rate, the units of measure are different and both units are stated, such as feet per second, miles per hour, and so on.

If a car travels 300 miles in 6 hours, we can find the rate by expressing the ratio of the measures 300 to 6 so that the second measure is 1.

$$\frac{300}{6} = \frac{50 \times 6}{1 \times 6} = \frac{50}{1} \times \frac{6}{6} = \frac{50}{1}$$

Since $\frac{50}{1} = 50$, we can state the rate as 50 miles per hour.

Exercises 9–3:

Find the ratio, in simplest form, of each pair of measurements below.

1. 8 oz. to 4 lb.   4. $.35 to $3.00
2. 12 gal. to 2 qt.   5. 9 in. to 3 ft.
3. 5 yd. to 18 in.   6. 32 hr. to 2 da.

Find the rate for each pair of measurements below.

7. 15° in 3 hr.   10. 250 gal. in 5 min.
8. $45 for 3 tons   11. 1200 mi. in 4 hr.
9. $75 in 5 hr.   12. 500 ft. in 5 sec.

PROPORTION

According to one cookbook 9 pounds of meat is needed to serve 15 people. If Mrs. Bedner uses this as a guide, how many pounds of meat are needed to serve 35 people?

We might think about this problem as follows: The ratio would remain the same regardless of how many people are to be served. If n represents the number of pounds of meat needed for 35 people, then $\frac{9}{15}$ and $\frac{n}{35}$ must name the same ratio.

$$\frac{9}{15} = \frac{n}{35}$$

This equation states the equality of two ratios. Such an equation is called a *proportion*.

Our knowledge of solving equations enables us to solve this proportion.

$$\frac{9}{15} = \frac{n}{35}$$

$$35 \times \frac{9}{15} = \frac{n}{35} \times 35 \quad \text{Mult. prop.}$$

$$\frac{35 \times 9}{15} = n$$

$$21 = n$$

Hence, 21 pounds of meat are needed to serve 35 people according to this particular cookbook.

Exercises 9–4:

Use a proportion to solve each of these problems.

1. Eight students eat 20 sandwiches at a party. How many sandwiches should you prepare for 12 students?

2. A 2-inch by 3-inch photograph was projected on a screen so that the width of the projection was 3 inches. What is the length of the projection?

3. In 5 hours Mr. Snowden drove 320 miles. At this rate, how many miles can he travel in 7 hours?

4. Twelve men can complete a job in 18 days. How many days will it take 18 men to complete the job?

5. A car can drive 72 miles on 5 gallons of gasoline. Under similar driving conditions, how many miles can this car travel on 22 gallons of gasoline?

6. An oil refinery produced 3 gallons of fuel oil for every 7 gallons of gasoline. How many gallons of fuel oil were produced while the refinery produced 126,000 gallons of gasoline?

7. A distance of 1 inch on a map represents an actual distance of 75 miles. The map distance between two cities is 6 inches. How many miles apart are the two cities?

8. A certain city has a population of 8000 inhabitants. Of 560 people chosen at random there were 140 under 18 years of age. Estimate the number of persons in this city less than 18 years of age.

TRANSLATING A PROBLEM INTO A PROPORTION

If an object travels 60 feet in 18 seconds, how long will it take to travel 70 feet (assuming constant speed)?

We can think about a proportion for this problem in several ways.

First, we may think of the ratio of each distance to its related time.

$$\frac{60}{18} = \frac{70}{t}$$

Second, we may think of each time to its related distance.

$$\frac{18}{60} = \frac{t}{70}$$

Third, we may think of the ratio of the first time to the second time and the first distance to the second distance.

$$\frac{18}{t} = \frac{60}{70}$$

Or you might think of the ratios of the second quantities to the first quantities.

$$\frac{t}{18} = \frac{70}{60}$$

Any of these four proportions is acceptable for the original problem. They are called equivalent proportions since they all have the same solution set, in this case {21}.

Exercises 9–5:

Use proportions to solve the following problems.

1. A 3-inch by 4-inch photograph was enlarged so that the width of the enlargement is 5 inches. How long is the enlargement?

2. While a jet plane flew 300 miles, a smaller propeller-driven plane flew 135 miles. If their speeds remain in this relationship, how many miles will the smaller plane fly while the jet flies 1000 miles?

3. A poll of 800 voters just before a local election revealed that 550 people intended to vote for

candidate A and 250 people intended to vote for candidate B.

If candidate A receives 3465 votes in the election, how many votes should candidate B expect to receive?

4. A librarian decided to order 3 fiction books for every 5 nonfiction books. She is making an order for 65 nonfiction books. How many fiction books should she order?

PER CENT

A ratio such as $\dfrac{7}{100}$ or $\dfrac{15}{100}$ in which the second number (or denominator) is 100 may also be expressed as a per cent.

Hence, 1 per cent means $\dfrac{1}{100}$ or .01.

In this sense, 15 per cent (denoted by 15%) means a ratio of 15 to 100 or a rate of 15 *per hundred*. Hence, per cent notation is still another way of naming a ratio and also a rational number.

$$19\% = \frac{19}{100} = .19$$

To change a ratio such as ⅗ to per cent notation, we can use the idea of a proportion where the second ratio has a denominator of 100.

$$\frac{3}{5} = \frac{n}{100}$$

$$100 \times \frac{3}{5} = \frac{n}{100} \times 100$$

$$60 = n$$

Therefore, $\dfrac{3}{5} = \dfrac{60}{100}$ or 60%.

State each of the following ratios as a per cent.

1. 18 to 50　　5. $\dfrac{17}{20}$　　9. $\dfrac{40}{100}$

2. 13 to 25　　6. $\dfrac{4}{5}$　　10. $\dfrac{300}{1000}$

3. 7 to 10　　7. $\dfrac{9}{20}$　　11. $\dfrac{17}{25}$

4. 3 to 20　　8. $\dfrac{1}{20}$　　12. $\dfrac{850}{1000}$

State each of the following as a decimal numeral.

13.	47%	16.	13%	19.	200%
14.	100%	17.	150%	20.	350%
15.	95%	18.	.6%	21.	1000%

USING PER CENTS

A transport truck delivered 360 gallons of gasoline to a service station. This was 30% of the total load. How many gallons of gasoline were on the truck before it delivered to that service station?

We can think about this problem as follows: 30 is to 100 as 360 is to some number n.

$$\frac{30}{100} = \frac{360}{n}$$

We already know this proportion can be stated as follows.

$$\frac{100}{30} = \frac{n}{360}$$

$$360 \times \frac{100}{30} = \frac{n}{360} \times 360$$

$$1200 = n$$

Therefore, the truck contained 1200 gallons before the delivery was made.

In the previous problem the per cent is sometimes referred to as the *rate*, the total number of gallons (n) as the *base*, and the number of gallons delivered as the *percentage*.

If we denote the *base* by b, the rate by r, and the percentage by p, most problems involving per cent can be solved by using the following formulas.

$$r = \frac{p}{b} \quad b = \frac{p}{r} \quad p = rb$$

There is little need to memorize all these formulas, since by remembering any one of them, and using the properties of solving equations, you can easily derive the others.

Since the rate is a per cent you can state it as a decimal or as a fraction in the computation. Study the following examples.

Example 1:
What number is 15% of 210?
$$p = rb$$
$$p = .15 \times 210$$
$$p = 31.5$$

Example 2:
45 is what per cent of 72?
$$45 = r \times 72$$
$$\frac{45}{72} = r$$
$$\frac{5}{8} = r$$
$$r = .625 \text{ or } 62.5\%$$

Example 3:
38 is 25% of what number?
$$p = rb$$
$$38 = .25b$$
$$\frac{38}{.25} = b$$
$$152 = b$$

Exercises 9–7:
Find the answer to each of these questions.
1. 28 is what per cent of 140?
2. What number is 18% of 56?
3. 32 is 40% of what number?
4. What number is 115% of 46?
5. 74 is what per cent of 296?
6. 114.7 is 58% of what number?

EQUIVALENT FRACTIONS AND PER CENTS

Since a per cent can be stated as a decimal or as a fraction, one of these forms is sometimes more convenient for computation than the other.

When a per cent such as 66⅔% is given, we can sometimes avoid approximate answers by using the fraction ⅔.

The following table gives the per cent, decimal, and fractional equivalents of some of the more common per cents.

Per Cent	Decimal	Fraction
50%	.50	$\frac{1}{2}$
$33\frac{1}{3}\%$	$.33\frac{1}{3}$	$\frac{1}{3}$
$66\frac{2}{3}\%$	$.66\frac{2}{3}$	$\frac{2}{3}$
25%	.25	$\frac{1}{4}$
75%	.75	$\frac{3}{4}$
20%	.20	$\frac{1}{5}$
$16\frac{2}{3}\%$	$.16\frac{2}{3}$	$\frac{1}{6}$
$83\frac{1}{3}\%$	$.83\frac{1}{3}$	$\frac{5}{6}$
$12\frac{1}{2}\%$	.125	$\frac{1}{8}$
$62\frac{1}{2}\%$	.625	$\frac{5}{8}$

To solve a problem that involves an equation such as

$$p = .75 \times 480$$

it is easier to replace .75 by ¾ and solve the following equation.

$$p = \frac{3}{4} \times 480 = 360$$

Exercises 9–8:

Solve the following problems.

1. A certain state has a retail sales tax of 4%. How many dollars in tax must be paid in that state when one purchases a new automobile for $3250?

2. Mr. Jacobs was dissatisfied with the gas mileage of 12 miles per gallon for his car. After having his car repaired, the gas mileage increased by 25%. What was his mileage then?

3. A club has a membership of 72 persons. Of these $16\frac{2}{3}\%$ are charter members. How many of the persons are charter members?

4. A news report stated that 70% of the families in Sunnydale own their own homes. If 560 families live in Sunnydale, how many own their own homes?

5. A quarterback completed 18 of 24 passes during a game. What per cent of his passes did he complete during the game?

SIMPLE INTEREST

When a person borrows money, the amount he borrows is called the *principal*. The money he pays for the use of money is called *interest*. The amount of interest depends on the rate (or per cent) of the principal he shall pay each year.

The interest formula is basically the same as the regular per cent formula, with one exception—the time is given as the number of years.

Let i = number of dollars of interest
p = number of dollars borrowed
t = number of years
r = rate

The simple interest formula is then given as follows.

$$i = p \times r \times t \text{ or } i = prt$$

Suppose you borrow $400 for 2 years at an annual (yearly) rate of 6%. The interest is computed as shown below.

$$i = prt$$
$$i = 400 \times .06 \times 2$$
$$i = 48$$

Therefore, at the end of 2 years you should pay $48 in interest.

Of course, the amount (denoted by a) you should repay is the principal plus the interest, or $400 + 48$ or $448.

$$a = p + i$$

If you want to compute the amount directly you might replace i in the above formula by prt (since $i = prt$), and use the following formula.

$$a = p + prt$$
$$a = p(1 + rt)$$

In case the money is borrowed for less than 1 year, we usually consider the year to be 12 months of equal length and consider a month to be 30 days in length. If the principal is used for 8 months, use $\frac{8}{12}$ or $\frac{2}{3}$ for t. If the principal is used for 90 days, use $\frac{90}{360}$ or $\frac{1}{4}$ for t.

Exercises 9–9:

Find the answer for each problem.

Principal	Rate	Interest	Time
1. $600	____	$27	9 mo.
2. ____	4%	$7.50	60 days
3. $1200	$3\frac{1}{2}\%$	____	$1\frac{1}{2}$ yr.
4. $900	____	$108	3 yr.
5. $400	6%	____	2 yr.

CHAPTER TEN

CONGRUENCE AND SIMILARITY

CONGRUENT LINE SEGMENTS

In the physical world we never expect two objects to have exactly the same length, but in our imaginary world of geometry every line segment is assumed to have many exact copies.

Given $\overline{AB}$ below, we could use a compass to construct $\overline{CD}$ so that the measures of the two line segments are the same.

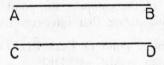

We do not say $\overline{AB} = \overline{CD}$ because they are obviously two different line segments, and $=$ means that two symbols name precisely the same thing.

Instead, we say that $\overline{AB}$ is congruent to $\overline{CD}$, or $\overline{AB} \cong \overline{CD}$.

Definition 10–1:
Two line segments are congruent if and only if they have the same measure. The symbol $\cong$ means *is congruent to*.

Exercises 10–1:
Answer these questions.
1. Can a horizontal line segment be congruent to a vertical line segment?
2. If $\overline{RS} \cong \overline{UV}$ and the measure of $\overline{RS}$ is 5½, what is the measure of $\overline{UV}$?
3. Are any two sides of a square congruent line segments?

4. If point K is the midpoint of $\overline{AB}$ is $\overline{AK} \cong \overline{KB}$?

BISECTING LINE SEGMENTS AND ANGLES

The word *bisect* means to separate into two parts of equal size. To bisect $\overline{AB}$ means to locate some point P on $\overline{AB}$ such that $\overline{AP} \cong \overline{PB}$.

To bisect $\overline{AB}$, open a compass so that the distance between the point of the compass and the tip of the pencil is greater than half the length of $\overline{AB}$. Place the compass tip at one end point of $\overline{AB}$ and then at the other to draw intersecting arcs as shown below.

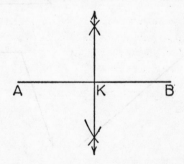

Draw the line determined by the points of intersection of the arcs. This line bisects $\overline{AB}$. Point K is called the *midpoint* of $\overline{AB}$, and $\overline{AK} \cong \overline{KB}$.

To bisect $\angle$ RST means to locate some point P in the interior of the angle so that m $\angle$ RSP $=$ m $\angle$ PST. (m stands for degree measure.)

To locate such a point, place the point of a compass at the vertex of the angle (point S in this case) and draw arcs that intersect the sides of the angle, as shown below.

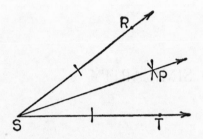

With these points as centers, draw two intersecting arcs in the interior of the angle. Call that point P. Draw $\overrightarrow{SP}$. Then m $\angle$ RSP = m $\angle$ PST and we say that $\angle$ RSP $\cong$ $\angle$ PST.

Definition 10–2:

Two angles are congruent if and only if they have the same measure.

Exercises 10–2:

Use a compass to bisect each line segment or angle below.

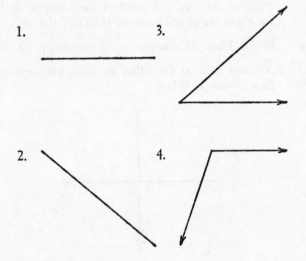

1.

3.

2.

4.

CONSTRUCTING CONGRUENT ANGLES

Given $\angle$ ABC below, how can we construct $\angle$ DEF so that the two angles are congruent?

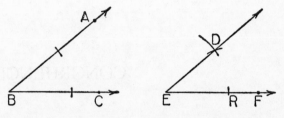

Draw ray EF. Place the point of a compass on vertex B and draw arcs that intersect the sides of $\angle$ ABC. Without changing the opening of the compass, use point E as a center and draw an arc as shown so that it intersects $\overrightarrow{EF}$ at some point R.

Now place the point of the compass at the point of intersection of one side of $\angle$ ABC and the arc. Open the compass so that the pencil tip is on the point of intersection of the other side of $\angle$ ABC and the arc. Without changing the opening of the compass, use point R as center and draw an arc that intersects the previously drawn arc at point D. Draw $\overrightarrow{ED}$. This construction makes $\angle$ ABC $\cong$ $\angle$ DEF.

Exercises 10–3:

Construct any angle congruent to each of the angles shown below.

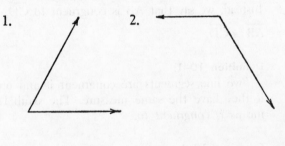

1.

2.

PERPENDICULAR LINES

Suppose we are given $\overleftrightarrow{AB}$ below and asked to construct some line RQ so that $\angle AQR \cong$ $\angle RQB$. We might consider $\overleftrightarrow{AB}$ as a straight angle whose vertex is point Q. Then we can proceed to bisect $\angle AQB$.

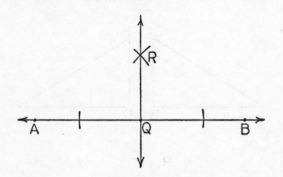

Since m $\angle AQB$ is 180 and since $\angle AQR \cong$ $\angle RQB$, then $\angle AQR$ is a right angle. Each of these newly formed angles has a degree measure of 90.

In this case we say that $\overleftrightarrow{AB}$ *is perpendicular to* $\overleftrightarrow{CD}$. This is denoted by $\overleftrightarrow{AB} \perp \overleftrightarrow{RQ}$.

Definition 10–3:

If two lines intersect so that right angles are formed, the lines are called *perpendicular lines*.

Suppose we are given $\overleftrightarrow{CD}$ and some point P not on $\overleftrightarrow{CD}$, and asked to construct a line through P that is perpendicular to $\overleftrightarrow{CD}$.

Hence, we must locate some point M on $\overleftrightarrow{CD}$ so that $\overleftrightarrow{PM} \perp \overleftrightarrow{CD}$. This can be done by doing the steps in the previous construction, but in the reverse order. Place the point of a compass on point P and draw arcs that intersect $\overleftrightarrow{CD}$ at some points called G and H, as shown below. Then bisect $\overline{GH}$; this is done by drawing the two arcs that intersect at J.

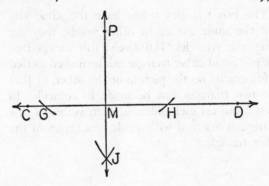

Draw $\overleftrightarrow{PJ}$ and label its intersection with $\overleftrightarrow{CD}$ with M. Since the angles formed by these two lines are all right angles, $\overleftrightarrow{CD} \perp \overleftrightarrow{PM}$.

Exercises 10–4:

Use a compass to construct the following.

1. A line perpendicular to $\overleftrightarrow{UV}$ at point P.

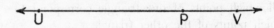

2. A line through point K perpendicular to $\overrightarrow{ST}$.

CONGRUENT TRIANGLES

We have learned that a general idea of congruence is that two figures are congruent if they have the same size and the same shape.

By drawing the necessary diagonals in any polygon we can form a figure composed entirely of triangles. Since triangles are so fundamental to much of mathematics and science, we need to know the conditions under which two triangles are congruent.

The two triangles below have the same size and the same shape. In other words, they are congruent triangles. Intuitively, this means that the picture of either triangle can be moved so that it fits exactly on the picture of the other, or that the two triangles can be made to coincide. In such a process each side and each vertex of one triangle is matched with a side or a vertex of the other triangle.

of sides or angles are called *corresponding parts* of the two triangles.

Exercises 10–5:

Assume that the two triangles in each exercise are congruent. For each pair of triangles, state the congruence between the triangles so that corresponding parts are indicated by the arrangements of the letters of the vertices.

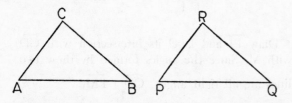

To be more specific about what we mean by saying that triangle ABC is congruent to triangle PQR, we have to tell which parts of the two triangles are matched or paired. In this way we tell which points of one triangle correspond to which points of the other triangle.

If we should move the picture of △ PQR onto the picture of △ ABC as follows, the two triangles will coincide.

Point P on point A
Point Q on point B
Point R on point C

We indicate this pairing in our notation of congruent triangles by arranging the letters of the vertices of each triangle so that the first letters correspond, the second letters correspond, and the third letters correspond.

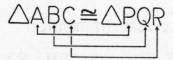

This statement is read *triangle ABC is congruent to triangle PQR.*

From this arrangement of the letters we also see that $\overline{AB}$ corresponds to $\overline{PQ}$, $\overline{BC}$ corresponds to $\overline{QR}$, and $\overline{CA}$ corresponds to $\overline{RP}$.

If two triangles are congruent, then for each angle or side of one triangle there is a congruent side or angle in the other triangle. Each such pair

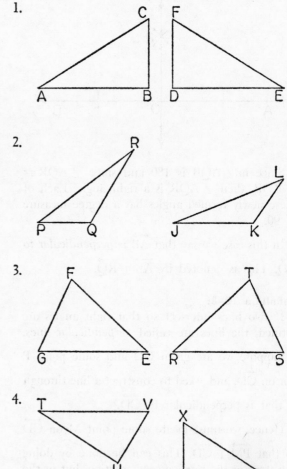

1.

2.

3.

4.

CONDITIONS FOR CONGRUENT TRIANGLES

Suppose we were given △ ABC below and asked to construct △ DEF so that △ ABC ≅ △ DEF.

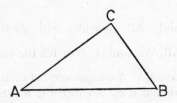

We could start by constructing $\overline{DE} \cong \overline{AB}$. This would locate two vertices of △ DEF.

We have three ways of locating the third vertex.

Case 1:

Construct ∠ DER ≅ ∠ ABC.

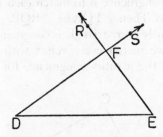

Then construct ∠ EDS ≅ ∠ BAC. The point of intersection of $\overrightarrow{ER}$ and $\overrightarrow{DS}$ is the third vertex.

In this case $\overline{DE}$ is part of both ∠ DEF and ∠ EDF, so let us refer to this case as *two angles and the included side.* Let us agree to accept the following condition for two triangles to be congruent.

Two triangles are congruent if two angles and the included side of one triangle are congruent respectively to two angles and the included side of the other triangle. Further, let us abbreviate this condition by A.S.A. (angle, side, angle).

Case 2:

Construct ∠ DET ≅ ∠ ABC.

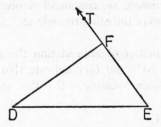

Then construct $\overline{EF} = \overline{BC}$. Then draw $\overline{DF}$.

This could also be done by constructing ∠ EDS ≅ ∠ BAC and $\overline{DF} \cong \overline{AC}$.

In this case, ∠ E is formed by sides DE and EF. We refer to this as *two sides and the included angle.* Let us agree to accept the following condition for two triangles to be congruent.

Two triangles are congruent if two sides and the included angle of one triangle are congruent respectively to two sides and the included angle of the other triangle (S.A.S.—side, angle, side).

Case 3:

Open a compass to the length of $\overline{BC}$. Use point E as center and draw an arc as shown below.

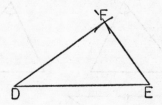

Now open the compass to the length of $\overline{AC}$. Use point D as center and draw an arc that intersects the previously drawn arc. This point of intersection is the third vertex F of the triangle, so draw $\overline{DF}$ and $\overline{EF}$.

Let us agree to accept the following condition for two triangles to be congruent.

Two triangles are congruent if the three sides of one triangle are congruent respectively to the three sides of the other triangle (S.S.S.—side, side, side).

As a matter of abbreviation the small marks on sides MN and GH indicate that these sides have the same measure or that they are congruent.

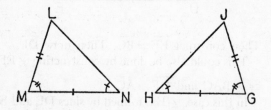

Also, the two marks on each of $\overline{ML}$ and $\overline{JG}$ indicate that they are congruent. We use the same idea for angles of two triangles. The marks above indicate that $\angle M \cong \angle G$ and that $\angle N \cong \angle H$.

Exercises 10–6:

Decide whether the pair of triangles in each exercise are congruent or not. If they are congruent, use A.S.A., S.A.S., or S.S.S. as a reason for your conclusion.

1.

2.

3.

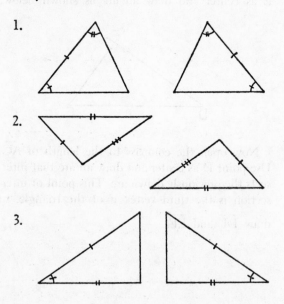

IDENTITY CONGRUENCE

Is every geometric figure congruent to itself? Consider $\overline{AB}$ below.

Certainly $\overline{AB}$ coincides with itself. That is, $\overline{AB} \cong \overline{AB}$. We could also match the end points so that $\overline{AB} \cong \overline{BA}$. A congruence between a geometric figure and itself is called the *identity congruence*.

Now consider $\angle$ PQR below.

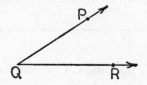

If we match each ray with itself we have $\angle$ PQR $\cong \angle$ PQR. Another way to establish the identity congruence is to match each ray with the other ray. Then $\angle$ PQR $\cong \angle$ RQP.

Obviously every triangle is congruent with itself only if we match every vertex with itself. This produces the identity congruence for triangles.

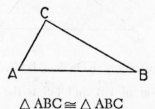

$$\triangle \text{ABC} \cong \triangle \text{ABC}$$

In order to establish a congruence between the two triangles shown below, we make use of the identity congruence.

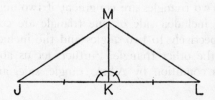

We are given the following.

$$\angle JKM \cong \angle LKM$$

$$\overline{JK} \cong \overline{LK}$$

By the identity congruence we know

$$\overline{MK} \cong \overline{MK}.$$

Therefore, $\triangle JKM \cong \triangle LKM$ because two sides and the included angle of one triangle are congruent to two sides and the included angle of the other triangle (S.A.S.).

Exercises 10–7:

In each figure below, congruent sides or angles are indicated by the small marks. Give four statements for each figure—three statements for congruence between corresponding parts and the fourth stating the congruence of two triangles and the condition (A.S.A., S.A.S., or S.S.S.) for this congruence.

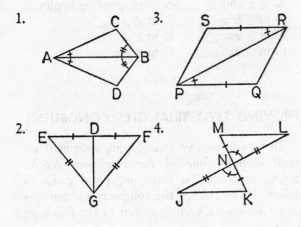

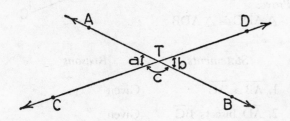

VERTICAL ANGLES

In the following figure, $\overleftrightarrow{AB}$ intersects $\overleftrightarrow{CD}$ at point T.

Recall that $\angle a$ and $\angle b$ are a pair of vertical angles.

$$m \angle a + m \angle c = 180$$
$$m \angle b + m \angle c = 180$$

Therefore, $m \angle a + m \angle c = m \angle b + m \angle c.$

Since $m \angle c$ names a number, we can use the addition property of equations to obtain the following.

$$m \angle a = m \angle b$$
$$\text{or}$$
$$\angle a \cong \angle c$$

A similar argument holds for every pair of vertical angles. We state this as: *vertical angles are congruent.*

PARALLEL LINES AND TRANSVERSALS

Parallel lines AB and CD below are intersected by line RS. Line RS is called a *transversal* of the parallel lines.

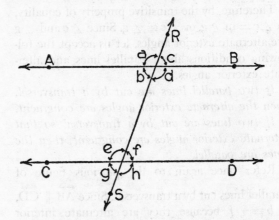

Since points A and D are on different lines and on opposite sides of the transversal, we can think of alternating from one side of the transversal to the other.

Furthermore, in the figure above, we refer to pairs of angles as follows:

$\angle b$ and $\angle f$ are *alternate interior angles*
$\angle d$ and $\angle e$ are alternate interior angles
$\angle a$ and $\angle h$ are *alternate exterior angles*
$\angle c$ and $\angle g$ are alternate exterior angles

Since $\angle c$ and $\angle f$ are in corresponding positions with regard to the two points of intersection, we can think of each pair of angles given below as a pair of *corresponding angles*.

$\angle c$ and $\angle f$ $\qquad$ $\angle b$ and $\angle g$
$\angle a$ and $\angle e$ $\qquad$ $\angle d$ and $\angle h$

Use a protractor to find the measure of $\angle b$ and the measure of $\angle f$. How do their measures compare? Do the same for $\angle d$ and $\angle e$.

Let us accept the following conditions about parallel lines and alternate interior angles.

If two parallel lines are cut by a transversal, then alternate interior angles are congruent.

If two lines are cut by a transversal so that alternate interior angles are congruent, then the lines are parallel.

Since $\overleftrightarrow{AB} \parallel \overleftrightarrow{CD}$ in the previous figure, then $\angle b \cong \angle f$. Furthermore, $\angle b \cong \angle c$ and $\angle f \cong \angle g$ because they are pairs of vertical angles. These statements of congruence can be translated into equalities of measures.

$$m \angle b = m \angle f$$
$$m \angle b = m \angle c$$
$$m \angle f = m \angle g$$

Therefore, by the transitive property of equality, $m \angle c = m \angle g$, or $\angle c \cong \angle g$. Since $\angle c$ and $\angle g$ are alternate exterior angles, let us accept the following conditions about parallel lines and alternate exterior angles.

If two parallel lines are cut by a transversal, then the alternate exterior angles are congruent.

If two lines are cut by a transversal so that alternate exterior angles are congruent, then the lines are parallel.

Refer once again to the previous figure of parallel lines cut by a transversal. Since $\overleftrightarrow{AB} \parallel \overleftrightarrow{CD}$, $\angle b \cong \angle f$ because they are alternate interior angles. Also, $\angle b \cong \angle c$ because they are vertical angles. Let us translate these statements into equalities of measures.

$$m \angle b = m \angle f$$
$$m \angle b = m \angle c$$

Therefore, $m \angle f = m \angle c$ or $\angle f \cong \angle c$. What kind of angles are $\angle f$ and $\angle c$?

Let us accept the following conditions about parallel lines and corresponding angles.

If two parallel lines are cut by a transversal, then corresponding angles are congruent.

If two lines are cut by a transversal so that corresponding angles are congruent, then the lines are parallel.

Exercises 10–8:

Use the figure below to complete the sentences that follow. Assume $\overleftrightarrow{JK} \parallel \overleftrightarrow{MN}$.

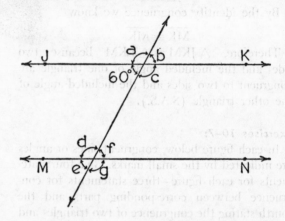

1. $\angle a$ and $\angle$ __ are alternate exterior angles.
2. $\angle c$ and $\angle$ __ are alternate interior angles.
3. $\angle g$ and $\angle$ __ are corresponding angles.
4. $m \angle b =$ __ $\qquad$ 7. $m \angle g =$ __
5. $m \angle e =$ __ $\qquad$ 8. $m \angle d =$ __
6. $m \angle c =$ __ $\qquad$ 9. $m \angle f =$ __

PROVING TWO TRIANGLES CONGRUENT

To prove that two triangles are congruent we must establish one of the conditions A.S.A., S.A.S., or S.S.S. To make our proof easier to follow, let us arrange the congruence statements and a reason for each as shown in the following example.

Given:

$\overline{AB} \cong \overline{AC}$

$\overline{AD}$ bisects $\overline{BC}$

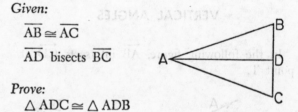

Prove:

$\triangle ADC \cong \triangle ADB$

Statements	Reasons
1. $\overline{AB} \cong \overline{AC}$	Given
2. $\overline{AD}$ bisects $\overline{BC}$	Given
3. $\overline{BD} \cong \overline{DC}$	Def. of bisect
4. $\overline{AD} \cong \overline{AD}$	Identity congruence
5. $\triangle ADC \cong \triangle ADB$	S.S.S.

Study the following example.

Given:

$\overline{QP} \cong \overline{RS}$

$\overleftrightarrow{QP} \parallel \overleftrightarrow{RS}$

Prove:

$\triangle RSQ \cong \triangle PQR$

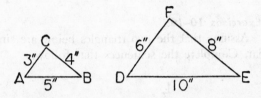

Statements	Reasons
1. $\overline{QP} \cong \overline{RS}$	Given
2. $\overleftrightarrow{QP} \parallel \overleftrightarrow{RS}$	Given
3. $\angle PQS \cong \angle RSQ$	Alt. int. angles
4. $\overline{QS} \cong \overline{QS}$	Identity congruence
5. $\triangle RSQ \cong \triangle PQR$	S.A.S.

Exercises 10–9:

Write the necessary statements and reasons to prove each of the following.

1. **Given:**

$\overrightarrow{CD}$ bisects $\angle ACB$

$\overline{AC} \cong \overline{BC}$

Prove:

$\triangle ADC \cong \triangle BDC$

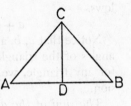

2. **Given:**

$\overleftrightarrow{MJ} \parallel \overleftrightarrow{KL}$

$\overleftrightarrow{ML} \parallel \overleftrightarrow{JK}$

Prove:

$\triangle JKL \cong \triangle LMJ$

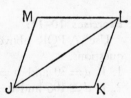

3. **Given:**

$\overline{SR}$ bisects $\overline{PQ}$

$\overline{PQ}$ bisects $\overline{RS}$

Prove:

$\triangle PTS \cong \triangle QTR$

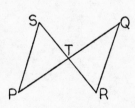

SIMILAR TRIANGLES

Generally speaking, any two geometric figures are similar if they have the same shape but not necessarily the same size. For example, we say that any two circles are similar, any two squares are similar, or any two line segments are similar.

Study the two triangles below. Do they appear to have the same shape? Are they the same size? Are they congruent?

In order that these two triangles have the same shape, the corresponding angles must be congruent. Just as for congruent triangles, let us arrange the letters of the vertices of two similar triangles so that they indicate the matching of corresponding vertices. For the above triangles we can say

$$\triangle ABC \sim \triangle DEF$$

and read this statement as *triangle ABC is similar to triangle DEF.*

Now let us find the ratios of the corresponding sides of these two triangles. Let us use the symbol

$\dfrac{\overline{AC}}{\overline{DF}}$ to denote the ratio of the measure of $\overline{AC}$ to the measure of $\overline{DF}$.

$$\frac{\overline{AC}}{\overline{DF}} = \frac{3}{6} = \frac{1}{2} \qquad \frac{\overline{AB}}{\overline{DE}} = \frac{5}{10} = \frac{1}{2} \qquad \frac{\overline{BC}}{\overline{EF}} = \frac{4}{8} = \frac{1}{2}$$

We notice that the same ratio exists for each pair of corresponding sides. We refer to this as *corresponding sides are proportional.*

Hence, we can conclude the following condition for two triangles to be similar.

Two triangles are similar if corresponding angles are congruent and corresponding sides are proportional.

Furthermore, *if two triangles are similar, then corresponding angles are congruent and corresponding sides are proportional.*

If you should construct two triangles so that corresponding angles are congruent, you will also find that corresponding sides are proportional. Furthermore, if you should construct two triangles such that corresponding sides are proportional, you will find that corresponding angles are congruent.

Hence, we do not need both conditions for two triangles to be similar. That is, *two triangles are similar if either the corresponding angles are congruent or the corresponding sides are proportional.*

Exercises 10–10:

Assume that the two triangles below are similar. Complete the sentences that follow.

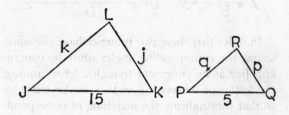

1. If m ∠ J = 40, then m ∠ P = __.

2. If m ∠ Q = 70, then m ∠ K = __.

3. If $k = 12$, then $q =$ __.

4. If $p = 2$, then $j =$ __.

5. In the figure below, $\overline{DE} \parallel \overline{AB}$. Give the necessary statements and reasons to show that △ ABC ~ △ DEC.

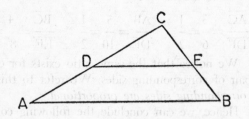

ANGLES OF A TRIANGLE

In the figure below, $\overleftrightarrow{AB}$ is drawn parallel to $\overleftrightarrow{PQ}$.

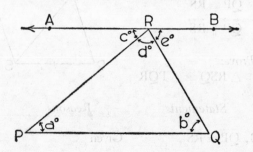

Since $\overleftrightarrow{AB}$ is a straight line, we know that $c + d + e = 180$.

Since $\overleftrightarrow{AB} \parallel \overleftrightarrow{PQ}$, we know that $a = c$ and $b = e$ because alternate interior angles are congruent when two parallel lines are cut by a transversal.

Since a and c name the same number, we can replace a by c in the sentence $c + d + e = 180$. Similarly, we can replace b by e since they name the same number. The resulting sentence is as follows.

$$a + b + d = 180$$

Notice that a, b, and d are the measures of the angles of the triangle. Since this argument holds for any triangle, we make the following conclusion.

The sum of the degree measures of the angles of a triangle is 180.

Exercises 10–11:

Use △ PQR above to answer the following questions.

1. If $a = 30$ and $b = 60$, then $d =$ __.

2. If $d = 106$ and $a = 24$, then $b =$ __.

3. If $a = b = d$, then the measure of each angle is __.

MORE ABOUT SIMILAR TRIANGLES

Assume in the two triangles below that $\angle$ A$\cong$ $\angle$ D and $\angle$ B $\cong$ $\angle$ E.

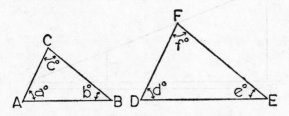

$$a+b+c=180 \qquad d+e+f=180$$
$$a+b+c=d+e+f$$

Since $\angle$ A $\cong$ $\angle$ D and $\angle$ B $\cong$ $\angle$ E, we know that $a=d$ and $b=e$. Hence, we can use the addition property of equations to deduce from $a+b+c=d+e+f$ that $c=f$ or $\angle$ C $\cong$ $\angle$ F.

Therefore, *if two angles of one triangle are congruent to two angles of another triangle, the third angles are congruent.*

USING SIMILAR TRIANGLES

To find the distance between points J and K on opposite banks of a river, Roger used a transit to measure the angles in forming the following figure.

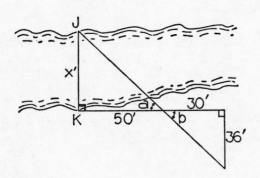

What kind of angles are a and b? Is $\angle$ a congruent to $\angle$ b? Since two angles of one triangle are congruent to two angles of the other, the two triangles are similar. Therefore, the following proportion holds.

$$\frac{x}{36}=\frac{50}{30}$$
$$x=\frac{50 \times 36}{30}$$
$$x=60$$

Therefore, the measure of $\overline{JK}$ is 60 feet.

Exercises 10–12:

Use proportional sides of similar triangles to solve these problems.

1. A lamppost 8 feet tall casts a shadow 12 feet long at the same time that a tree casts a shadow 42 feet long. How tall is the tree?

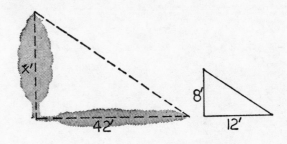

2. To find the distance between points P and Q at opposite ends of a pond, a transit was used to measure the angles to form this figure. What is the length of $\overline{PQ}$?

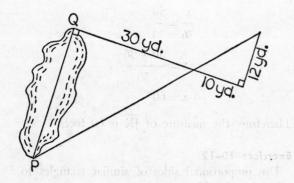

3. The structure of a bridge formed the triangles shown below. Find the length of $\overline{AB}$.

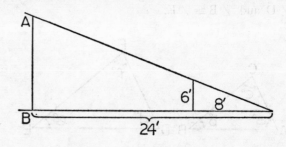

CHAPTER ELEVEN

PERIMETER, AREA, VOLUME

PERIMETER OF A RECTANGLE

The word *perimeter* means the distance around or the measurement of the boundary of a figure. Since the perimeter is a measurement, it is stated by a *numeral* and a *unit* of measure.

Every simple closed figure has a perimeter. To find the perimeter of any polygon, we can add the measures of its sides, assuming that all measures are determined by using the same unit of measure.

Definition 11–1:

A *rectangle* is a quadrilateral (four-sided polygon) in which opposite sides are parallel and all angles are right angles.

Let us investigate another property of any rectangle ABCD as shown below.

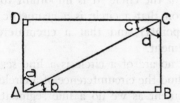

Knowing that it is a rectangle is sufficient for us to conclude that $\overline{DC} \parallel \overline{AB}$ and $\angle D \cong \angle B$. Diagonal AC is a transversal of both pairs of parallel sides. Alternate interior angles are congruent when parallel lines are cut by a transversal. Therefore, $\angle a \cong \angle d$ and $\angle b \cong \angle c$. By the identity congruence, $\overline{AC} \cong \overline{AC}$. Thus, $\triangle ABC \cong \triangle CDA$ (A.S.A.). Hence $\overline{BC} \cong \overline{AD}$ and $\overline{AB} \cong$

$\overline{DC}$ because they are corresponding parts of congruent triangles.

$\overline{BC}$ and $\overline{AD}$ are opposite sides of the rectangle.

Similarly, $\overline{AB}$ and $\overline{DC}$ are opposite sides. We have proved that opposite sides of a rectangle are congruent.

Knowing that opposite sides of a rectangle are congruent helps us in finding its perimeter. For example, consider rectangle PQRS below.

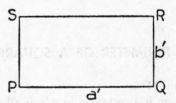

If the measure of $\overline{PQ}$ is a', then the measure of $\overline{SR}$ is also a'. If the measure of $\overline{RQ}$ is b', then the measure of $\overline{SP}$ is also b'.

To find the perimeter of rectangle PQRS we can add the measures of its sides. Let us use the letter p to represent the number of feet in its perimeter.

$$p = a + a + b + b$$
$$= 2a + 2b$$
$$= 2(a + b)$$

In other words, to find the perimeter of a rectangle we can find the sum of the measures of two adjacent sides and then multiply this sum by two.

Any two adjacent sides of a rectangle are commonly referred to as the *length* and the *width* of the rectangle, and their measurements are referred to as the dimensions of the rectangle.

To find the perimeter of a rectangle whose length is 16 inches and whose width is 7 inches (or whose dimensions are 16 in. and 7 in.) we can proceed as follows. Let l represent the measure of the length and let w represent the measure of the width.

$$p = 2(l + w)$$

Then replace l by 16 and w by 7.

$$\begin{aligned} p &= 2(l + w) \\ &= 2(16 + 7) \\ &= 2 \times 23 \\ &= 46 \end{aligned}$$

The perimeter is 46 inches.

Exercises 11–1:

Find the perimeter of a rectangle that has the following dimensions.

1. 12 ft. by 18 ft.
2. 54 in. by 32 in.
3. 5.7 in. by 3.8 in.
4. 3½ yd. by 7½ yd.
5. 2.03 in. by 9.89 in.
6. 129 ft. by 854 ft.

PERIMETER OF A SQUARE

Definition 11–2:

A *square* is a rectangle in which all four sides are congruent to each other.

To find the perimeter of a square each side of which is 8 inches long, we merely add the measures.

$$\begin{aligned} p &= 8 + 8 + 8 + 8 \\ &= 4 \times 8 \\ &= 32 \end{aligned}$$

We see that adding the measures of the sides is equivalent to multiplying the measure of any one of the sides by four.

If the measurement of a side of a square is s, then we can find the perimeter according to the following.

$$\begin{aligned} p &= s + s + s + s \\ &= 4s \end{aligned}$$

Exercises 11–2:

Find the perimeter of a square that has sides of the following length.

1. 9 in.
2. 15 ft.
3. 1.8 ft.
4. 13 in.
5. 3½ yd.
6. 1.26 in.

Find the length of a side of a square that has the following perimeter.

7. 28 in.
8. 80 ft.
9. 116 in.
10. 72 yd.
11. 9.2 ft.
12. 4.96 in.

CIRCUMFERENCE OF A CIRCLE

We have already defined a circle as a set of points in a plane each of which is the same distance from some point in the plane called the center of the circle.

Definition 11–3:

A line segment joining the center of a circle with any point on the circle is called a *radius*. The plural of radius is *radii*.

Definition 11–4:

A line segment that joins two points of a circle and passes through the center of the circle is called a *diameter*.

The perimeter of a circle is called the *circumference* of the circle. It is important to note the distinction that a circle is a geometric figure (a set of points) and that a circumference is a measurement.

Since no arc of a circle is a line segment, we cannot find the circumference of a circle by direct measurement as we do a line segment.

One method of approximating the circumference is explained below. Mark some point—call it B—on a circular object such as a can lid or a circular paper plate. As shown below, place point B at some point R on a line and roll the object along the line until point B is again on the line. Let S name this new position of point B on the line.

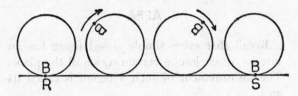

By measuring $\overline{RS}$ you can obtain an approximation of the circumference of the circle. We might also find the measure of the diameter of the circle and compare the measure of the circumference to the measure of the diameter. The following table gives a few such comparisons, where C represents the measure of the circumference and d represents the measure of the diameter of the same circle.

Diameter	Circumference	$\dfrac{C}{d}$
1.19 in.	3.75 in.	$\dfrac{3.75}{1.19} = 3.16$
2.50 in.	7.81 in.	$\dfrac{7.81}{2.50} = 3.13$
6.00 in.	18.81 in.	$\dfrac{18.81}{6.00} = 3.14$

In each case the result is between 3.10 and 3.20. Mathematicians know that $\dfrac{C}{d}$ is the same for all circles regardless of their size. Experimentation is not the only way to approximate this number. It has been computed to several thousand decimal places and never repeats nor terminates. To the nearest 4 decimal places it is 3.1416; however, for most purposes 3.14 is precise enough.

Since the number obtained by $\dfrac{C}{d}$ cannot be named by a fraction and is always the same, it is referred to by the Greek letter *pi* (denoted by π) and pronounced "pie."

The following statements summarize our findings.

$$\frac{C}{d} = \pi \quad \text{and} \quad C = \pi d$$

In other words, π is the ratio of C to d, and to find the circumference of a circle we find the product of π and the measure of its diameter.

If the radius of a circle is 4.2 inches long, what is its circumference?

$$\begin{aligned} C &= \pi d \\ &= \pi \times 4.2 \\ &= 3.14 \times 4.2 \\ &= 13.188 \end{aligned}$$

Using our rule for finding the product of two approximate numbers we would state the result with only 2 significant digits. That is, we would say the circumference is 13 inches.

Exercises 11–3:

Answer the following.

1. If a radius of a circle is 3 inches long, how long is the diameter of the circle?

2. If a diameter of the circle is 8.6 feet long, how long is a radius?

3. How many radii does a circle have?

4. How many diameters does a circle have?

Use 3.14 for π and find the circumference of a circle having the following measurement of a radius or a diameter.

5. diameter, 3.2 in. 8. radius, 4.3 in.
6. diameter, 6.0 ft. 9. radius, 12 ft.
7. diameter, 24 in. 10. radius, 15 yd.

PERIMETER OF OTHER CLOSED FIGURES

We can now find the perimeter of any closed figure whose boundary is composed of line segments or arcs of a circle.

Find the perimeter of the figure below.

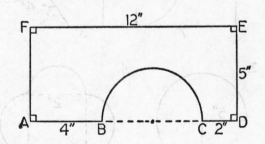

Since all of the angles are right angles we conclude that the basic figure is a rectangle. Hence, $\overline{AF}$ is 5 in. long. All we need do now is to find the length of the half-circle and add the measures.

Since $\overline{AD}$ is 12 in. long we can see that $\overline{BC}$ is 6 in. long and is the diameter of the circle.

The length of $\overparen{BC}$ is ½ of the circumference.

$$\text{measure of } \overparen{BC} = \frac{1}{2} \ (\pi d)$$

$$= \frac{1}{2} \ (3.14 \times 6)$$

$$= \frac{1}{2} \times 18.84$$

$$= 9.42$$

The perimeter of the figure is then the sum of the following measures, naming the parts of the boundary clockwise from point A.

$$p = 5 + 12 + 5 + 2 + 9.42 + 4$$
$$= 37.42$$

Since the sum of the measures can be no more precise than the least precise measure involved, we should state the perimeter to the nearest inch. Hence, the perimeter is 37 inches.

Exercises 11–4:

Find the perimeter of each figure

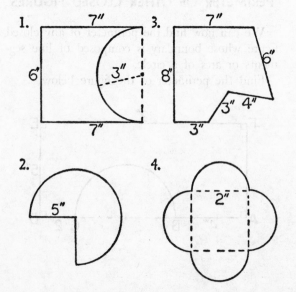

AREA

Recall that every simple closed figure has an interior, or encloses a certain region of the plane. The measurement of such a region is called its *area.*

Certainly a rectangle or a triangle has zero area (or no area) since it is merely the set of points forming the figure. However, we usually use a phrase such as "the area of the triangle" to mean the area of the region it encloses.

To find the area of a simple closed figure we use a procedure similar to that of finding the measure of an angle or a line segment. First, we arbitrarily choose some region as our unit of measure. It is most convenient to choose the interior of a square as our unit of area measure. As shown below, a square with sides 1 inch long has an area of 1 *square inch*; a square with sides 1 foot long has an area of 1 *square foot.*

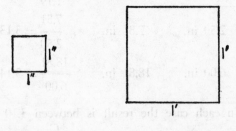

Area is 1 sq. in.   Area is 1 sq. ft.

AREA OF A RECTANGLE

To find the area of the rectangle below we can proceed as follows.

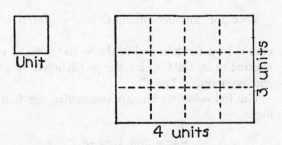

Use the side of the unit square to determine the measure of each side of the rectangle. The measurement 4 units tells us how many unit

squares fit along that side, and the measurement 3 units tells us how many such rows of unit squares are needed to completely cover the interior of the rectangle. The dotted lines in the figure above separate the interior into squares each of which is equivalent in area to the unit of area measure.

We could count the number of these squares to determine that the area of the rectangle is 12 square units. But we see that this is merely an application of multiplication. Let A represent the number of square units in the area.

$$A = 4 \times 3$$

This is but a specific case. Suppose the length of a rectangle is l units and the width is w units. Then we can express the area by the following sentence.

$$A = lw$$

Suppose a rectangle is 15″ long and 7″ wide. We can find its area by replacing l by 15 and w by 7.

$$\begin{aligned} A &= lw \\ &= 15 \times 7 \\ &= 105 \end{aligned}$$

The area is 105 square inches.

We could call this the computed area since it is a result of computation without applying the rule for multiplying measures. By applying the rule for multiplying measures, we should round 105 to one significant digit and state the area as 100 square inches.

Exercises 11–5:

State the area of a rectangle of the following dimensions in two ways: the computed area and the area obtained by applying the rule for multiplying measures.

1. 2 ft. 8 ft.
2. 12 in. by 11 in.
3. 9 yd. by 22 yd.
4. 3.2″ by 8.5″
5. .75′ by .43′
6. 8.12″ by 3.24″

AREA OF A SQUARE

Since a square is a special kind of rectangle, we can also use

$$A = lw$$

to find the area of a square.

Suppose each side of a square is s units long. What is its area?

$$\begin{aligned} A &= lw \\ &= s \times s \\ &= s^2 \end{aligned}$$

In other words, to find the area of a square, we square the measure of one of its sides. For example, a square with each side 7 inches long has an area of 49 square inches.

$$\begin{aligned} A &= s^2 \\ &= 7^2 \\ &= 49 \end{aligned}$$

Applying the rule for operating with measures, we should round 49 to 50 so that we have only one significant digit and state the area as 50 square inches.

Exercises 11–6:

State the area of a square having the following dimensions in two ways: the computed area and the area obtained by using the rule for multiplying measures.

1. Each side 5 in.
2. Each side 8 in.
3. Each side 17 ft.
4. Each side 4.3 in.
5. Each side 11.5 ft.
6. Each side 21 yd.

AREA OF A RIGHT TRIANGLE

Suppose we are given the right triangle shown below and asked to find its area.

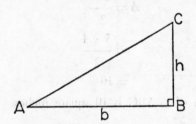

Let us refer to $\overline{AB}$ as the *base* of the triangle and denote its measure by b. Let us refer to $\overline{BC}$ as the height (or altitude) of the triangle and denote its measure by h. Of course, there is nothing wrong with interchanging these.

Then we could think of constructing a line through C parallel to $\overleftrightarrow{AB}$ and of constructing a line through A parallel to $\overleftrightarrow{BC}$ as shown below. Call their point of intersection point D.

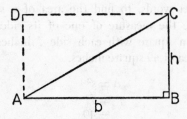

Since ABCD is a rectangle and $\overline{AC}$ is a diagonal of the rectangle, we can prove that $\triangle ABC \cong \triangle CDA$. Thus, the two triangles have the same size, which means that they have the same area. In other words, the area of either triangle is one-half the area of the rectangle.

Area of rectangle: bh

Area of $\triangle ABC$: $\dfrac{1}{2}bh$ or $\dfrac{bh}{2}$

Therefore, if A represents the number of units in the area of $\triangle ABC$, then

$$A = \frac{bh}{2}.$$

A similar construction can be done for any right triangle, so the above relationship holds for all right triangles.

If $b = 5$ and $h = 4$ in $\triangle ABC$, then we find its area as follows.

$$A = \frac{bh}{2}$$
$$= \frac{5 \times 4}{2}$$
$$= 10$$

The area of $\triangle ABC$ is 10 square inches.

Exercises 11–7:

Find the computed area of a right triangle having these dimensions.

	Base	Height	Area
1.	14 in.	8 in.	___ sq. in.
2.	25 ft.	12 ft.	___ sq. ft.
3.	5.7 in.	8.2 in.	___ sq. in.
4.	42 yd.	36 yd.	___ sq. yd.

AREA OF A TRIANGLE

Since not all triangles are right triangles, we are confronted with many situations in which we are to find the area of a triangle that is not a right triangle, such as the triangle below.

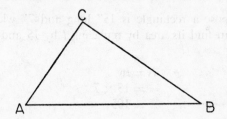

Once again, let us use what we already know in developing a way to find the area of $\triangle ABC$.

Construct a line through C perpendicular to $\overline{AB}$ as shown below.

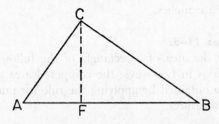

Line segment CF is called an altitude of $\triangle ABC$.

Definition 11–5:

An *altitude* of a triangle is the perpendicular line segment joining any vertex to the line that contains the opposite side of the triangle.

Sometimes an altitude of a triangle (except for its end points) is in the interior of the triangle, sometimes an altitude is a side of the triangle, and sometimes an altitude (except for one end point) is in the exterior of the triangle. In each figure below $\overline{AB}$ is an altitude.

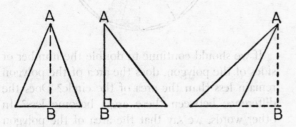

Furthermore, every triangle has three different altitudes, as shown below. $\overline{AB}$, $\overline{CD}$, and $\overline{EF}$ are altitudes.

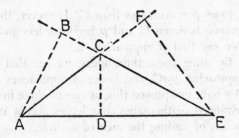

Now let us return to the problem of finding the area of $\triangle ABC$.

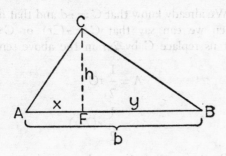

Area of $\triangle AFC = \dfrac{xh}{2}$

Area of $\triangle BFC = \dfrac{yh}{2}$

Since the area of $\triangle ABC$ is equal to the sum of the areas of $\triangle AFC$ and $\triangle BFC$, let us add their area measures.

$$\text{Area of } \triangle ABC = \frac{xh}{2} + \frac{yh}{2}$$

$$= \frac{xh + yh}{2}$$

$$= \frac{(x+y)h}{2}$$

$$= \frac{bh}{2}$$

We can state this in words as: *The area of a triangle is one-half the product of the length of any base and the length of the altitude to that base.*

Suppose in the previous figure that the length of $\overline{AB}$ is 8 inches and the length of $\overline{CF}$ is 5 inches.

$$A = \frac{bh}{2}$$

$$= \frac{8 \times 5}{2}$$

$$= 20$$

The area is 20 square inches.

Exercises 11–8:

Find the computed area of a triangle having these dimensions.

	Base	Altitude	Area
1.	18 in.	10 in.	___ sq. in.
2.	52 ft.	13 ft.	___ sq. ft.
3.	5.6 in.	7.3 in.	___ sq. in.
4.	28 yd.	15 yd.	___ sq. yd.

AREA OF A CIRCLE

Just as with rectangles and triangles, we refer to the area of a circular region as the area of the circle. Since no part of a circle is a line segment, the problem of finding the area of a circle is more difficult than for polygons.

To get an idea of the area of the circle below, let us pick points on the circle so that the line segments joining any two consecutive points are congruent.

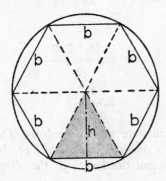

Since all radii of a circle have the same length, they are congruent. We have selected points on the circle so that all line segments labeled b are congruent. Are all of these triangles congruent? Why?

The area of any one of these triangles is $\frac{1}{2} bh$.

Then the area of the polygon is 6 times as great.

$$A = \left[\frac{1}{2} bh\right] 6$$

$$= \left[\frac{1}{2} h\right] (b \times 6) \quad \text{Why?}$$

$$= \left[\frac{1}{2} h\right] (6b) \quad \text{Why?}$$

Since $6b$ represents the perimeter of the polygon, we can express the area of the polygon as

$$A = \frac{1}{2} hp$$

where p represents the number of units in the perimeter of the polygon.

Certainly the area of the circle is greater than the area of the polygon. We can make the dif-

ference between these areas less and less by doubling the number of sides of the polygon.

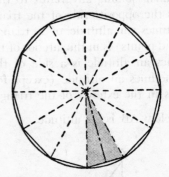

If we should continue to double the number of sides of the polygon, does the area of the polygon remain less than the area of the circle? Does the difference between these areas become less? In other words, we say that the area of the polygon approaches the area of the circle.

Does h remain less than r (the measure of the radius)? However, the difference between r and h becomes less and less. We say that h approaches r.

Does p remain less than C? However, the difference between C and p becomes less and less. We say that p approaches C.

By using these three ideas, we see that $\frac{1}{2}hp$ approaches $\frac{1}{2}rC$, and hence A approaches $\frac{1}{2}rC$. We have not proved this, as can be done in more advanced mathematics, but let us accept it as a means of finding the area of a circle.

$$A = \frac{1}{2} rC$$

We already know that $C = \pi d$ and that $d = 2r$. Then we can say that $C = \pi(2r)$ or $C = 2\pi r$. Let us replace C by $2\pi r$ in the above sentence.

$$A = \frac{1}{2} rC$$

$$= \frac{1}{2} r(2\pi r)$$

By the commutative and associative properties of multiplication this becomes

$$A = \frac{1}{2} \times 2 \times \pi \times r \times r$$

$$= \pi r^2$$

In other words, to find the area of a circle, find the product of π and the square of the measure of the radius.

Suppose we are to find the area of a circle whose radius is 5 inches long.

$$A = \pi r^2$$
$$= 3.14 \times 5^2$$
$$= 3.14 \times 25$$
$$= 78.5$$

The computed area is 78.5 square inches.

Exercises 11–9:

Find the computed area of a circle whose radius has the following length.

1. 2 in. 3. 12 yd. 5. 25 in.
2. 4 ft. 4. 3.2 in. 6. 18.7 ft.

Solve these problems.

7. Mr. Benson made a circular flower garden with a radius 8 feet long. How many feet of fence are needed to enclose it? What is the area of the garden?

8. A radio station installed a new transmitter that enables it to broadcast a distance of 70 miles. If you consider the surface of the earth near the station to be part of a plane, how many square miles are there in the station's broadcasting area?

PRISMS

In the figure below, think of moving $\triangle$ CDE from plane m to a parallel plane n so that $\overline{ET}$, $\overline{CR}$, and $\overline{DS}$ are parallel line segments.

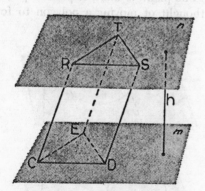

The resulting three-dimensional figure is called a *prism*. The triangular region CDE is called the *base* of the prism and the triangular region RST is called the *upper base* of the prism.

Since the bases are congruent triangles, this figure is called a *triangular prism*. Line segments ET, CR, and DS are called lateral edges of the prism. A perpendicular line segment between the two planes, indicated by h in the figure, is called an *altitude* of the prism.

Any other kind of polygon can be used as the base. If a rectangle is used as the base, as shown below, the prism is called a *rectangular prism*.

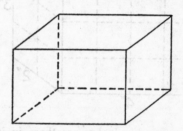

We shall consider only those prisms in which the lateral edges are perpendicular to both bases. Such prisms are called *right prisms* (since perpendicular implies right angles). The rectangular prism shown above is a *right rectangular prism*. This means that the bases are congruent rectangles and that the lateral edges are perpendicular to both bases.

VOLUME OF A RIGHT PRISM

A special kind of right rectangular prism has all of its edges the same length. That is, the bases are congruent squares and each lateral edge has the same length as the side of the base. Such a prism is called a *cube*.

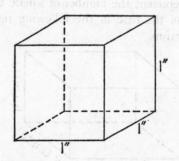

In this cube each edge is 1 inch long. Its interior contains 1 *cubic inch*. If each edge of a cube is 1 foot long, its interior contains 1 *cubic foot*.

The measurement of the interior of a prism is called its *volume*.

The interior of the following right rectangular prism is separated into cubes each having a volume of 1 cubic inch.

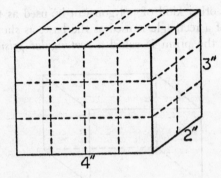

How many cubic inches are there in its interior? What is its volume?

We see in the drawing that there are 3 layers of cubes with 4×2 cubes in each layer. It appears that we can find the volume by multiplying the number of cubes in each layer by the number of layers. Let V represent the number of cubic inches in the volume.

$$V = (4 \times 2) \times 3$$
$$= 8 \times 3$$
$$= 24$$

The volume is 24 cubic inches.

How would you find the area of the base of the prism? Is the number of square inches in the area of the base the same as the number of inch cubes in each layer? Could you multiply the area of the base by the measure of the altitude of the prism to find its volume?

Let B represent the number of square units in the area of the base in the following right rectangular prism.

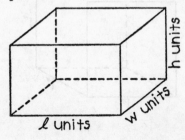

$$V = lwh \text{ or } V = Bh$$

We can use either of these formulas, whichever is more convenient, to find the volume of a prism. Suppose a box has the shape of a right rectangular prism. It is 8 inches long, 6 inches wide, and 5 inches high. What is its volume?

$$V = lwh$$
$$= 8 \times 6 \times 5$$
$$= 240$$

Its volume is 240 cubic inches.

Exercises 11–10:

Find the volume of right rectangular prisms that have the following dimensions.

	Length	Width	Altitude
1.	10 in.	4 in.	7 in.
2.	24 ft.	13 ft.	8 ft.
3.	3.2 in.	1.8 in.	5.6 in.
4.	9 yd.	7 yd.	6 yd.
5.	2½ ft.	1¾ ft.	4 ft.

Solve these problems.

6. How many cubic yards of earth will be moved in making an excavation that is 50 feet long, 26 feet wide, and 9 feet deep?

7. A steel tank has the shape of a right rectangular prism. Its dimensions are 3 feet, 2 feet, and 6 feet. How many cubic feet are in its volume?

8. If water weighs 62.5 pounds per cubic foot, what is the weight of the water that can be placed in the tank described in exercise 7?

CYLINDERS

In the following figure, think of moving a circle from one plane to another parallel plane just as you thought of moving a polygon to form a prism.

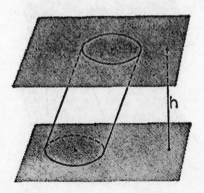

The resulting figure is called a *cylinder*. In fact, this is called a *circular cylinder* since both of its bases are circles.

There are other kinds of cylinders depending upon the curved figure used as a base. However, we shall be concerned only with circular cylinders. Furthermore, we shall discuss only right circular cylinders—circular cylinders in which the line segment joining the centers of the two circular bases is perpendicular to both bases.

Examples of a right circular cylinder are common vegetable or fruit cans or oil barrels.

VOLUME OF A RIGHT CIRCULAR CYLINDER

How can we find the volume of the right circular cylinder below?

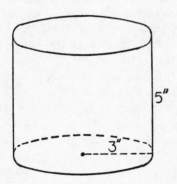

We could find the area of its base. Let B represent the number of square inches in the area of the base.

$$B = \pi r^2$$
$$= 3.14 \times 3^2$$
$$= 3.14 \times 9$$
$$= 28.26$$

Just as with a prism, this means that a layer containing 28 inch cubes, to the nearest cube,

would cover the region of the base. The measure of the altitude then tells us how many such layers are needed to completely fill the cylinder.

It appears that we can find the volume of a cylinder by multiplying the area measure of its base by the measure of its altitude. This means that the formula

$$V = Bh$$

is also usable for finding the volume of a cylinder.

Thus, the volume of the right circular cylinder above is found as follows:

$$V = Bh$$
$$= \pi r^2 h$$
$$= 3.14 \times 3^2 \times 5$$
$$= 3.14 \times 9 \times 5$$
$$= 141.3$$

Its volume to the nearest cubic inch is 141 cubic inches.

Exercises 11–11:

Find the volume of right circular cylinders having the following dimensions.

	Radius of Base	Altitude
1.	5 in.	7 in.
2.	2 ft.	3 ft.
3.	1.6 in.	4.2 in.
4.	3 in.	78 in.
5.	1 yd.	5 yd.

Solve these problems.

6. A cylindrical can is 6 inches high and the radius of its base is 3 inches. How many cubic inches are there in its volume?

7. A rock of irregular shape was placed in a right cylindrical container. Water was added until the rock was completely submerged in the water. When the rock was removed, the level of the water lowered 3 inches. If the radius of the container was 5 inches long, find the volume of the rock.

VOLUME OF A RIGHT CIRCULAR CYLINDER

CHAPTER TWELVE

PROBABILITY AND STATISTICS

PROBABILITY

If you heard the weatherman predict, "There is a chance for rain tomorrow," would you be likely to wear your raincoat tomorrow?

Suppose the weatherman said, "It will probably rain tomorrow." Would you be likely to wear your raincoat?

Suppose he said, "The chance of rain tomorrow is about 7 out of 10." Would you be likely to wear your raincoat?

Another way of saying the "chance" that an event will occur is to say the "probability" that the event will occur. For example, there are 2 red cards and 1 blue card in a box. Without looking, you reach in and select one of the cards. What is the probability of selecting a red card?

There are 3 possible outcomes—you may select any one of the cards. Furthermore, each of the outcomes is equally likely. This means that the chance of drawing one particular card is the same as for any other card.

Since we are interested in selecting a red card, let us call selecting a red card a successful outcome. Since there are 2 red cards in the box, there are 2 possible successful outcomes. Hence, we say the probability of selecting a red card is 2 out of 3. We also denote "2 out of 3" by the fraction $\frac{2}{3}$.

Definition 12–1:

Let n represent the total number of possible outcomes of an event. Let s represent the number of successful outcomes. The *probability* of a successful outcome is $\frac{s}{n}$.

A bowl contains 3 red marbles and 5 black marbles. A marble is drawn at random without looking. What is the probability that the marble is red?

$$n = 8 \text{ and } s = 3$$

The probability is $\frac{3}{8}$.
In other words, you have 3 chances out of 8 of drawing a red marble.

What is the probability of drawing a black marble?

$$n = 8 \text{ and } s = 5$$

The probability is $\frac{5}{8}$.
In other words, you have 5 chances out of 8 of drawing a black marble.

Exercises 12–1:

A box contains 11 slips of paper. The slips are numbered 0 through 10. A slip is drawn sight unseen. What is the probability that the numeral on the slip drawn names:
1. The number seven
2. An even number
3. A number less than six
4. A number greater than eight
5. An odd number
6. The number thirteen

A bowl contains 2 red marbles, 5 black marbles, and 7 blue marbles. One marble is drawn. What is the probability that it will be (state the fraction in lowest terms):
7. Red 11. Not black
8. Black 12. Not blue
9. Blue 13. Neither red nor blue
10. Not red 14. Neither black nor red

PROBABILITY OF SUCCESS OR FAILURE

A hat contains 8 red cards and 4 green cards. You are blindfolded and draw one card from the hat. What is the probability that it is a red card?

In this case $n = 12$ and $s = 8$. The probability is $\frac{8}{12}$ or $\frac{2}{3}$.

What is the probability of *not* drawing a red card? That is, what is the probability that the outcome is *not* successful?

In this case the probability is $\frac{4}{12}$ or $\frac{1}{3}$.

What is the sum of the probability that a successful outcome occurs and the probability that a successful outcome will *not* occur?

$$\frac{2}{3} + \frac{1}{3} = \frac{2+1}{3} = \frac{3}{3} = 1$$

We can express this as follows: If P represents the probability that a successful outcome occurs and F represents the probability that a failure (not a successful outcome) occurs, then
$$P + F = 1.$$
We can apply the properties of equations and state this as $P = 1 - F$ or $F = 1 - P$.

For example, if the probability of the White Sox winning the pennant is $\frac{2}{5}$, then their probability of not winning the pennant is $1 - \frac{2}{5}$ or $\frac{3}{5}$.

Exercises 12–2:

A cup contains 3 pennies, 2 nickels, and 4 dimes. The cup is shaken until one coin falls out. Any coin is likely to fall out. Find the probability that the first coin to fall is:

1. A penny
2. Not a penny
3. A dime
4. Not a dime
5. A nickel
6. Not a nickel

You close your eyes and press down one of the 26 letter keys on a typewriter. Find the probability that this key prints:

7. The letter Q
8. Not the letter Q
9. A vowel
10. Not a vowel
11. A letter in the name *Jane*
12. Not a letter in the name *Jane*
13. A letter that comes before G
14. Not a letter that comes before G

MORE THAN ONE OUTCOME

A bag contains 15 marbles. Four of these are green, two are black, and the rest have other colors. If you draw one marble, what is the probability it will be either green *or* black?

To be successful you can draw either a green marble *or* a black marble. There are 6 successful outcomes and 15 possible outcomes. Therefore, the probability is $\frac{6}{15}$ or $\frac{2}{5}$.

On the other hand, the probability of drawing a green marble is $\frac{4}{15}$. The probability of drawing a black one is $\frac{2}{15}$. To find the probability of a successful outcome we can add the probabilities of these two outcomes.

The probability of a green marble is $\frac{4}{15}$.

The probability of a black marble is $\frac{2}{15}$. The probability of a green marble *or* a black marble is

$$\frac{4}{15} + \frac{2}{15} = \frac{4+2}{15} = \frac{6}{15} = \frac{2}{5}.$$

Exercises 12–3:

You close your eyes and press down one of the 26 letter keys on a typewriter. Find the probability that the letter you print is:

1. C or D
2. A letter in the name *Rudy* or *Jean*
3. A vowel or the letter T
4. A letter before D or a letter after W
5. A letter between K and Q or a letter between S and V

TWO OUTCOMES IN SUCCESSION

Suppose that a bowl contains one cutout of each of the letters A, B, and C. You draw one letter and replace it. The letters are then mixed and you draw another letter. What is the probability that you draw the letter C both times?

We might investigate all of the possible outcomes for the two draws. Let us give the letter of the first draw first, and the letter of the second draw second. For example, suppose you draw A on the first draw and C on the second draw. We can denote this by (A,C). The set of all possible outcomes can be listed as follows.

$$\left\{ \begin{array}{ccc} (A,A), & (A,B), & (A,C) \\ (B,A), & (B,B), & (B,C) \\ (C,A), & (C,B), & (C,C) \end{array} \right\}$$

How many of these 9 possible outcomes denotes that C was drawn both times? Therefore, the probability of drawing twice in succession is $\frac{1}{9}$.

Now let us investigate the problem in another way. Let us consider the probabilities of the individual draw. What is the probability of drawing C on the first draw? What is the probability of drawing C on the second draw?

If we find the product of these probabilities, we obtain the same probability as in the first investigation.

$$\frac{1}{3} \times \frac{1}{3} = \frac{1}{9}$$

In other words, when the objects of a set are replaced after each outcome, the probability of successive outcomes is the product of the individual probabilities.

If you toss a coin it must show either a head or a tail. Each outcome is equally likely. Suppose you toss a coin twice. What is the probability of getting 2 heads?

The following denotes all of the possible outcomes.

$$\{(T,T), (T,H), (H,T), (H,H)\}$$

The probability of getting 2 heads is $\frac{1}{4}$.

If we disregard the order and ask, "What is the probability of getting a head and a tail?" how would you find the probability? Could you add the probabilities of (T,H) and (H,T) to obtain this probability?

Furthermore, the individual probabilities do not have to be the same to find the probability of successive outcomes. Suppose the probability of success on the first outcome is $\frac{2}{3}$ and of success on the second outcome is $\frac{1}{4}$; then the probability of these two outcomes occurring in succession is their product.

$$\frac{2}{3} \times \frac{1}{4} = \frac{2 \times 1}{3 \times 4} = \frac{2}{12} = \frac{1}{6}$$

Exercises 12–4:

There are 3 red cards and 7 green cards in a box. One card is drawn and replaced. A second card is drawn. Find the probability that:

1. Both cards are red
2. Both cards are green
3. The first card is red and second card is green.

Two men, Al and Ed, are playing a game. The probability that Al will win is $\frac{4}{9}$. The probability that Ed will win is $\frac{2}{9}$. The probability that the game will end in a tie is $\frac{1}{3}$. Find the probability that:

4. Al wins two games in succession
5. First Ed wins, then Al wins
6. Ed wins two games in succession
7. First Ed wins, than a tie game
8. First a tie game, then Al wins
9. Two tie games in succession
10. In two games each man wins one game (Caution: The order of winning is not considered.)

MORE ABOUT PROBABILITY

Now let us consider the case where replacements are not made. That is, the first object drawn is *not* replaced in the set.

For example, there are 5 red hats and 3 white hats in a closet. You walk into the dark closet, select a hat, come out and place the hat on a table. Then you repeat this to get a second hat. What is the probability that you selected 2 red hats?

For the first outcome there are 8 hats, 5 of which are red. The probability of the first hat being red is $\frac{5}{8}$.

However, for the second outcome there are only 7 hats remaining in the closet and only 4 of these are red. Therefore, the probability of the second hat being red is $\frac{4}{7}$.

To find the probability of both hats being red, find the product of the individual probabilities.

$$\frac{5}{8} \times \frac{4}{7} = \frac{5 \times 4}{8 \times 7} = \frac{20}{56} = \frac{5}{14}$$

To find the probability of the first hat being white and the second hat being red, first find the individual probabilities.

First hat white: ⅜

Second hat red: 5/7

Then find the product of these individual probabilities.

$$\frac{3}{8} \times \frac{5}{7} = \frac{15}{56}$$

Therefore, the probability of first getting a white hat and then a red one is ¹⁵⁄₅₆, which means 15 chances out of 56.

Exercises 12–5:

There are 3 red cards and 7 green cards in a box. One card is drawn and *not* replaced. Then a second card is drawn. Find the probability that:

1. Both cards are red
2. Both cards are green
3. The first card is red and the second card is green

There are 12 cards lying face down on a table. Each card is marked with a different letter and 5 of the letters are vowels. You turn over two cards in succession. After the first card is turned over, it is left face up. Find the probability that you turn up:

4. Two vowels
5. Two letters that are not vowels
6. First a vowel and then not a vowel
7. First not a vowel and then a vowel

You are to draw a card from a regular 52-card deck of playing cards. Do *not* replace this card. Then draw a second card. What is the probability that the cards drawn are:

8. Both aces
9. First a spade and then a heart
10. First a king and then a queen
11. First a king or a queen and then an ace

TOSSING COINS

An interesting study in probability is the investigation of tossing coins. We have already mentioned the probabilities of various combinations of heads or tails when a single coin is tossed twice. That is, on the first toss the coin shows either a head or a tail. A similar situation holds for the second toss. We can list the various outcomes as a set of ordered pairs.

$$\{(H,H), (H,T), (T,H), (T,T)\}$$

There are 4 possible outcomes and each outcome is equally likely. The probability of any one of these outcomes is ¼. If we disregard the order in which the coins show a head or a tail, and simply ask for the probability that the outcome is a head and a tail, we see that (H,T) and (T,H) both denote this outcome. Hence, the probability of this outcome is 2/4 or ½.

Now let us consider the case of tossing two coins at the same time. In this case we are not concerned about which coin falls first or second, but only the possible outcomes. Let us refer to the coins as Coin A and Coin B. We can obtain a listing of the possible outcomes as follows.

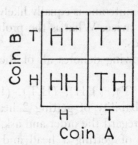

There are 4 possible outcomes. One outcome is HH, two outcomes are an H and a T, and one outcome is TT. We see that the following probabilities exist.

2 heads: $\dfrac{1}{4}$

1 head, 1 tail: $\dfrac{2}{4} = \dfrac{1}{2}$

2 tails $\dfrac{1}{4}$

We see that these are the same probabilities that we obtained by tossing a single coin twice. In other words, the possible outcomes and their probabilities are the same whether we toss a single coin twice or toss two coins only once.

What are the various possible outcomes and their probabilities of occurring if we should toss 3 coins only once? Certainly the outcomes of any 2 of these 3 coins will be the same as those found above. Then each of these outcomes will be matched with an H and then with a T. That is, there would be twice as many possible outcomes as there were for 2 coins. The set of possible outcomes is shown below.

$$\left\{ \begin{array}{l} (H,H,H),\ (H,H,T),\ (H,T,H),\ (H,T,T) \\ (T,H,H),\ (T,H,T),\ (T,T,H),\ (T,T,T) \end{array} \right\}$$

There are 8 possible outcomes. The following probabilities exist.

3 heads: $\dfrac{1}{8}$

2 heads, 1 tail: $\dfrac{3}{8}$

1 head, 2 tails: $\dfrac{3}{8}$

3 tails: $\dfrac{1}{8}$

How many possible outcomes are there when 1 coin is tossed only once?

How many possible outcomes are there when 2 coins are tossed only once?

How many possible outcomes are there when 3 coins are tossed only once?

Let us list these numbers of coins and the numbers of outcomes in a table to help discover an important pattern.

Number of coins	1	2	3	4	...	n
Number of outcomes	2	4	8	?	...	2^n

That is, if we toss 3 coins there are 2^3 or 8 possible outcomes; if we toss 4 coins there are 2^4 or 16 possible outcomes; and so on. This idea enables us to determine n in $\dfrac{s}{n}$ when finding the probability of a particular outcome.

Can we discover an easy way to determine s in $\dfrac{s}{n}$ when finding the probability of a particular outcome? Let us investigate the probabilities of the outcomes already investigated. In particular, let us observe the various values of s (or numerators of the probabilities).

For 1 coin:

$$\{H,T\}$$
$$1\quad 1$$

For 2 coins:

$$\{\ (H,H),\ (H,T),\ (T,H),\ (T,T)\ \}$$
$$1\qquad\qquad 2\qquad\qquad 1$$

For 3 coins:

$$\left\{ \begin{array}{c} (H,H,H),\ (H,H,T),\ (H,T,T),\ (T,T,T) \\ (H,T,H),\ (T,H,T) \\ (T,H,H),\ (T,T,H) \end{array} \right\}$$
$$1\qquad\quad 3\qquad\quad 3\qquad\quad 1$$

We are not in a position to prove that this pattern continues, but it can be proved in more advanced mathematics. These values of s can be arranged in a triangular arrangement or array known as Pascal's triangle, an important array in probability theory.

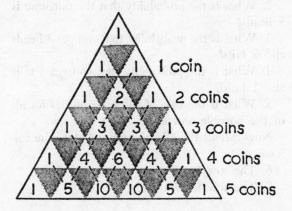

Notice in this array that to go from the numbers in one row to the numbers in the next row we need only find sums of any two consecutive numbers.

How can we determine probabilities by using Pascal's triangle?

(*a*) The numbers in any row are the values of

$$\frac{s}{n}$$

s in –.

(*b*) The sum of the numbers in any row is

the value of *n* in $\frac{s}{n}$.

Which combination of heads and tails is represented by each number in a particular row? Let us illustrate this by considering 4 coins being tossed at the same time. The number of possible outcomes is $1+4+6+4+1$ or 16.

Combination	Number of Outcomes	Probability
4 heads	1	$\frac{1}{16}$
3 heads, 1 tail	4	$\frac{4}{16}$ or $\frac{1}{4}$
2 heads, 2 tails	6	$\frac{6}{16}$ or $\frac{3}{8}$
1 head, 3 tails	4	$\frac{4}{16}$ or $\frac{1}{4}$
4 tails	1	$\frac{1}{16}$

Exercises 12–6:

Answer these questions about tossing 5 coins at the same time.

1. How many possible outcomes are there?

2. What is the probability that the outcome is 5 heads?

3. What is the probability that you get 3 heads and 2 tails?

4. What is the probability that you get 4 tails and 1 head?

5. What is the sum of the probabilities for all of the possible outcomes?

Now extend Pascal's triangle and state the following.

6. The row for 6 coins

7. The row for 7 coins

MEASURES OF CENTRAL TENDENCY

The following is a set of test scores obtained by a class of 19 students.

82, 65, 48, 72, 78, 92, 76, 86, 86, 68, 54, 72, 81, 86, 90, 83, 86, 61, 40

By just looking at the scores as they are listed we aren't able to tell very much about the class.

To get a better idea of the class, let us list the scores in numerical order from least to greatest.

40, 48, 54, 61, 65, 68, 72, 72, 76, 78, 81, 82, 83, 86, 86, 86, 86, 90, 92

Can we give a single measure or score which is most representative of this set of scores? That is, can we give a single score that reveals something about the class? Such a measure is called a *measure of central tendency*. It describes a measure about which the others seem to cluster or centralize.

We shall briefly discuss three measures of central tendency.

THE MEAN OR AVERAGE

A person might be tempted to find the average score. That is, find the sum of all of the scores and divide by the number of scores in the set.

For the previous set of test scores the sum of the scores is 1406. There are 19 scores in the set.

Hence the average is $\frac{1406}{19}$ or 74.

This type of average is called the *arithmetic mean* or simply the *mean*. The mean is usually considered the most useful measure of central tendency and is undoubtedly the most common. It is easily expressed as a formula and takes into account every number in the set. However, if extreme values occur, the mean may lead us to a false impression of the set of numbers or scores.

Exercises 12–7:

Find the mean for each set of numbers named below.

1. {7,2,8,3,4,6}
2. {120,130,50,60,90}
3. {9,5,12,13,4,5,9,7}
4. {10,12,13,15,10,5,5}

THE MODE

The number that occurs most frequently in a set of numbers is another measure of central tendency, called the *mode*. If no number occurs more than once, the mode does not exist (there is no mode). If two or more numbers occur the same number of times there can be more than one mode for the set of numbers.

For example, the mode of this set of numbers {1,1,3,5,7,19} is 1 since it occurs twice and is the only number that occurs more than once.

The set of numbers {5,11,17,32,46} has no mode since each number occurs only once.

The set of numbers {2,2,5,7,7,11,27} has two modes, 2 and 7, since each of these numbers occur twice and no other number occurs more than twice.

The mode is often thought to be the most representative since it occurs most frequently. It is easy to determine, but there may be no mode at all or more than one mode. Also, not every number is taken into account when finding the mode.

Exercises 12–8:
Find the mode or modes of each set of numbers named below.

1. {81,85,85,89,90,91,95}
2. {41,41,48,10,12,12,16,12}
3. {60,52,63,57,49,61}

THE MEDIAN

When numbers are arranged in their natural order, either least to greatest or greatest to least, the *median* is the number which is in the middle position. There are just as many numbers above the median as there are below the median.

The median of the set of numbers {4,5,6,7,8} is the middle number, which is the number 6.

The median of the set of numbers {9,7,5,4,3,2,1} is the number 4.

When a set contains an even number of members, there is no middle number. In this case we usually give the mean of the two middle numbers as the median. For example, for the set of numbers {17,19,23,25,28,31} the two middle numbers are 23 and 25. The mean of 23 and 25 is 24. Therefore, the median of this set of numbers is 24.

The median is not influenced by extreme values. However, if values are clustered in distinct and widely separated groups, the median may lead to a poor impression of the set of numbers as a whole.

Exercises 12–9:
Find the median of each set of numbers named below.

1. {50,65,65,67,70,77,84}
2. {125,136,300,136,210,200,286}
3. {66,72,85,87,90,97}
Solve these problems.
4. Suppose you own a store and sell different brands of soap as shown.

Brand A	4 bars
Brand B	9 bars
Brand C	14 bars
Brand D	30 bars

If you should continue to stock only one of these brands, which should you select? Which measure of central tendency is used?

5. The following prices for 4-ounce bottles of hair tonic are obtained from seven different manufacturers.

30¢, 30¢, 32¢, 34¢, 46¢, 48¢, 53¢

Which measure of central tendency is most descriptive of the set of prices? What is this price?

6. The high temperature readings in degrees Fahrenheit for the first day of each month in the town of Buffalo Run were recorded as follows.

January	12	July	105
February	12	August	105
March	32	September	62
April	55	October	45
May	76	November	17
June	105	December	12

Find the mean and the median temperature readings. Do these measures indicate that Buffalo Run has a pleasant year-round temperature? Now find the modes of this set of temperature readings. Now do you think that Buffalo Run has a pleasant year-round temperature?

ANSWERS TO EXERCISES

Ex. 1–1:
1. {Huron, Superior, Erie, Michigan, Ontario}
2. {October, November, December}
3. {Texas, Louisiana, Mississippi, Alabama, Florida}
4. $\emptyset$ or the empty set
5. {January, February, March, April, May, June, July, August, September, October, November, December}
6. {Alaska, Alabama, Arizona, Arkansas}
7. The set of the first 4 letters of the English alphabet
8. The set of vowels in the English alphabet
9. The set of the last 3 letters of the English alphabet

10. $\in$ 12. $\notin$ 14. $\notin$
11. $\in$ 13. $\notin$ 15. $\in$

Ex. 1–2:
1. $\not\subset$ 3. $\subset$ 5. $\subseteq$
2. $\not\subset$ 4. $\subset$ 6. $\subseteq$

7. {x,y}, {x}, {y}, $\emptyset$
8. {a,b,c,d}, {a,b,c}, {a,b,d}, {a,c,d}, {b,c,d}, {a,b}, {a,c}, {a,d}, {b,c}, {b,d}, {c,d}, {a}, {b}, {c}, {d}, $\emptyset$
9. If a set contains n members, then it contains 2^n subsets.

Ex. 1–3:
1. $\neq$ 3. $\neq$ 5. $\neq$ 7. $=$ 9. $\neq$
2. $=$ 4. $=$ 6. $\neq$ 8. $\neq$ 10. $\neq$

Ex. 1–4:
Answers may vary.
1. {a, b, c, d}
   | | | |
   {w, x, y, z}
2. {1, 2, 3, 4, 5, 6}
   | | | | | |
   {2, 4, 6, 8, 10, 12}

Ex. 1–5:
1. 8   3. 9   5. 0   7. 36
2. 7   4. 12   6. 4   8. 50

Ex. 1–6:
1. 85   3. 730   5. 3432
2. 539   4. 777   6. 6051

7. $(4 \times 10) + (6 \times 1)$
8. $(1 \times 100) + (2 \times 10) + (4 \times 1)$
9. $(6 \times 100) + (2 \times 10) + (9 \times 1)$
10. $(8 \times 10) + (2 \times 1)$
11. $(3 \times 1000) + (4 \times 100) + (2 \times 10) + (6 \times 1)$
12. $(2 \times 1000) + (0 \times 100) + (4 \times 10) + (1 \times 1)$

Ex. 1–7:
1. 10^5 3. 10^4 5. 4^6
2. 10^2 4. 7^4

6. $10 \times 10 \times 10$
7. $10 \times 10 \times 10 \times 10 \times 10$
8. $10 \times 10 \times 10 \times 10 \times 10 \times 10 \times 10$
9. $6 \times 6 \times 6 \times 6$

Ex. 1–8:
1. 1,100,002,826   3.   712,309
2.   5,000,001   4. 52,000,018

Ex. 1–9:
1. 29,000;   28,600;   28,560
2. 71,000;   70,800;   70,840
3. 53,000;   53,100;   53,150

Ex. 2–1:
1. $J \cup K = \{1,2,3,5,7,9\}$
2. $K \cup M = \{2,3,4,5,6,7,8,9\}$
3. $K \cup N = \{0,3,5,7,9\}$
4. $J \cup M = \{1,2,3,4,6,8\}$
5. $J \cup N = \{0,1,2,3,5,9\}$
6. $M \cup N = \{0,2,4,5,6,8,9\}$
7. $N \cup M = \{0,2,4,5,6,8,9\}$
8. $K \cup K = K$

Ex. 2–2:
1. $C \cap D = \{2,3\}$ 5. $D \cap E = \{3\}$
2. $D \cap C = \{2,3\}$ 6. $D \cap F = \{7,8\}$
3. $C \cap E = E$ 7. $E \cap F = \emptyset$
4. $C \cap F = \emptyset$ 8. $E \cap E = E$

Ex. 2–3:
1. T   4. T   7. T   10. T
2. T   5. T   8. T
3. F   6. T   9. F

Ex. 2–4:
1. 11, 10, 12, 14, 13
2.   7, 12, 14, 10, 18
3. 16, 10, 16,   9, 15
4. 15, 13, 12, 11, 13
5. 11, 10, 10, 14, 12

Ex. 2–5:
1. No   3. No   5. No
2. Yes   4. No   6. No

7. $3 + 7 = 7 + 3$
8. $8 + 15 = 15 + 8$
9. $36 + 17 = 17 + 36$
10. $156 + 13 = 13 + 156$
11. $129 + 47 = 47 + 129$
12. $326 + 218 = 218 + 326$
13. $327 + 56 = 56 + 327$
14. $651 + 87 = 87 + 651$

Ex. 2–6:
1. Yes   2. Yes   3. Yes

4. $5 + (7 + 6) = (5 + 7) + 6$
5. $17 + (15 + 32) = (17 + 15) + 32$
6. $(9 + 8) + 7 = 9 + (8 + 7)$
7. $13 + (12 + 6) = (13 + 12) + 6$
8. $(72 + 31) + 46 = 72 + (31 + 46)$

9. 16   10. 27   11. 13   12. 29

Ex. 2–7:
1. A   4. C, A   7. A   10. A
2. C   5. C   8. C
3. C   6. C, A   9. C, A

Ex. 2–8:
1. $4 + 6 = 10$ 2. $8 + 3 = 11$ 3. $2 + 6 = 8$

Ex. 2–9:
1. 16, 17, 24, 132, 109
2. 20, 20, 40, 29, 150

Ex. 2–10:
1. 878, 7999, 59998
2. 777, 8847, 88978
3. 689, 1979, 37927
4. 368, 6788, 18796

Ex. 2–11:
1. 6021, 79870,   908182
2. 8021,   58322, 1006033
3. 4923, 194967,   674437
4. 2267,   44007,   330068

Ex. 2–12:
1. Open your eyes
2. Sit down (or lie down)
3. Return from school
4. Open your book
5. Take 5 steps backward
6. Tie your shoe
7. Subtract seven
8. Add thirteen

Ex. 2–13:
1. 15 3. 69 5. 756 7. 312
2. 754 4. 26 6. 39 8. r

9. 8 12. 6 15. 7
10. 3 13. 9 16. 9
11. 8 14. 7

Ex. 2–14:
1. 6 3. 8 5. 9
2. 4 4. 8 6. 9

7. $9 + n = 12$, 3 awards
8. $15 - n = 7$, 8 cupcakes
9. $12 - 8 = n$, 4 children
10. $7 + n = 13$, 6 apples

Ex. 2–15:
1. $10 - 4 = 6$ 2. $9 - 6 = 3$ 3. $9 - 7 = 2$

Ex. 2–16:
1. 225, 5234, 46323
2. 113, 3013, 20344
3. 621, 3135, 34711

Ex. 2–17:
1. 152, 1318, 34882
2. 182, 1907, 26151
3. 188, 2387, 26803

Ex. 2–18:
1. 163   2109   21467
   +152   +1318   +34882
   ----   -----   ------
    315   3427   56349

2. 226   3475   20858
   +182   +1907   +26151
   ----   -----   ------
    408   5382   47009

3. 537   3856   43197
   +188   +2387   +26803
   ----   -----   ------
    725   6243   70000

Ex. 3–1:
1. 9 2. 8 3. 6 4. 24

Ex. 3–2:
1. 15 4. 14 7. 24 10. 28
2. 24 5. 16 8. 30 11. 18
3. 10 6. 16 9. 27 12. 9

Ex. 3–3:
1. $2 + 2 + 2 + 2 + 2 + 2$; 12
2. $1 + 1 + 1 + 1 + 1 + 1$; 6
3. 6; 6
4. $5 + 5 + 5 + 5$; 20
5. $4 + 4 + 4 + 4 + 4$; 20

6. $7 + 7 + 7$; 21
7. $7 + 7 + 7 + 7 + 7 + 7$; 42
8. $9 + 9 + 9 + 9$; 36
9. $7 + 7 + 7 + 7 + 7$; 35

10. $4 \times 8 = 32$ 13. $5 \times 9 = 45$
11. $6 \times 5 = 30$ 14. $2 \times 9 = 18$
12. $5 \times 1 = 5$ 15. $9 \times 2 = 18$

Ex. 3–4:
1. $5 \times 9 = 9 \times 5$ 6. $23 \times 5 = 5 \times 23$
2. $7 \times 8 = 8 \times 7$ 7. $9 \times 18 = 18 \times 9$
3. $31 \times 7 = 7 \times 31$ 8. $27 \times 13 = 13 \times 27$
4. $12 \times 6 = 6 \times 12$ 9. $357 \times 6 = 6 \times 357$
5. $9 \times 17 = 17 \times 9$ 10. $43 \times 127 = 127 \times 43$

Ex. 3–5:
1. $(3 \times 7) \times 5 = 3 \times (7 \times 5)$
2. $4 \times (2 \times 3) = (4 \times 2) \times 3$
3. $(6 \times 3) \times 2 = 6 \times (3 \times 2)$
4. $17 \times (8 \times 5) = (17 \times 8) \times 5$
5. $(13 \times 9) \times 3 = 13 \times (9 \times 3)$

6. A 8. C   10. C   12. A
7. C 9. C, A 11. C, A

Ex. 3–6:
1. $7 \times (2 + 5) = (7 \times 2) + (7 \times 5)$
2. $4 \times (3 + 6) = (4 \times 3) + (4 \times 6)$
3. $(4 + 5) \times 3 = (4 \times 3) + (5 \times 3)$
4. $(8 + 9) \times 6 = (8 \times 6) + (9 \times 6)$
5. $(5 \times 2) + (7 \times 2) = (5 + 7) \times 2$
6. $(6 \times 3) + (8 \times 3) = (6 + 8) \times 3$

Ex. 3–7:
1. 60   4. 800 7. 400 10. 4000
2. 900   5. 200 8. 4000
3. 30   6. 40   9. 400

Ex. 3–8:
1. 210   3. 4500 5. 24000
2. 3200 4. 1400   6. 140000

Ex. 3–9:

1. 168   4. 2124   7. 188   10. 2568
2. 390   5. 296   8. 145   11. 1371
3. 744   6. 3834   9. 1038   12. 5706

Ex. 3–10:

1. 494   4. 1296   7. 4067
2. 648   5. 4056   8. 3876
3. 1645 6. 986   9. 2325

Ex. 3–11:

1. 1872   4. 134784   7. 13704
2. 12768 5. 83025   8. 144672
3. 10044 6. 273132   9. 1101562

Ex 3–12:

1. $1200 < n < 2100$, 1755
2. $5600 < n < 7200$, 6150
3. $1500 < n < 2400$, 2262
4. $3600 < n < 5000$, 4042
5. $12000 < n < 20000$, 17249
6. $28000 < n < 40000$, 37884
7. $40000 < n < 54000$, 42952
8. $30000 < n < 42000$, 38324

Ex 3–13:

1. 4   4. 4   7. 2   10. 2
2. 3   5. 5   8. 4   11. 8
3. 3   6. 9   9. 3   12. 6

Ex. 3–14:

1. 6   5. 9   9. 5
2. 7   6. 7   10. 7
3. 6   7. 5   11. 9
4. 9   8. 4   12. 4

Ex. 3–15:

1. 8   4. 11   7. 13
2. 13 5. 12   8. 12
3. 13 6. 6   9. 19

Ex. 3–16:

1. 3 r2   4. 6 r1   7. 13 r3
2. 3 r3   5. 4 r4   8. 9 r1
3. 2 r1   6. 5 r3   9. 11 r1

Ex. 3–17:

1. 15   4. 63   7. 415
2. 26   5. 72   8. 421
3. 34   6. 67   9. 352

Ex. 3–18:

1. 16 r15   4. 7 r13   7. 123 r0
2. 17 r7   5. 17 r0   8. 124 r7
3. 11 r11   6. 12 r2   9. 200 r0

Ex. 3–19:

1. 6 r13   5. 123 r25   9. 322 r9
2. 38 r0   6. 201 r17   10. 24 r0
3. 42 r3   7. 137 r6   11. 102 r0
4. 10 r22 8. 211 r9   12. 86 r0

Ex. 4–1:

1. −7   3. 0   5. 0
2. −15   4. −12 6. −306

Ex. 4–2:

1. 4   4. 14   7. 3
2. 8   5. 22   8. 12
3. 5   6. 35   9. 23

Ex. 4–3:

1. −5   4. −13   7. −57   10. −122
2. −12   5. −19   8. −59
3. −12   6. −35   9. −128

Ex. 4–4:

1. −3   5. 21   9. −12   13. −17
2. −9   6. 12   10. 104 14. −71
3. −6   7. −2   11. 157
4. −3   8. 2   12. −78

Ex. 4–5:

1. 11   5. −21   9. −31
2. −20   6. −5   10. −113
3. 4   7. 5   11. 257
4. −9   8. 21   12. −12

Ex. 4–6:

1. −12   5. −120   9. −2247
2. −16   6. −120   10. −1025
3. −24   7. −104   11. −750
4. −35   8. −273   12. −2898

Ex. 4–7:

1. 15   5. 28   9. 120   13. −1608
2. 28   6. 120   10. −105 14. 630
3. −28   7. −120   11. −366
4. −28   8. −120   12. 366

Ex. 4–8:

1. 0   5. —315   9. 144
2. 54   6. 0   10. —900
3. —1066   7. —624   11. 90
4. 0   8. —108   12. 1

Ex. 4–9:

1. —5   5. —9   9. —25
2. —2   6. —9   10. —25
3. —8   7. 5   11. —32
4. —9   8. 25   12. —12

Ex. 4–10:

1. 17   5. 7   9. 12   13. —9
2. 1   6. —12 10. —13 14. 9
3. —1   7. 12   11. 13
4. 7   8. —12 12. 9

Ex. 5–1:

1. F   4. T   7. T   10. F
2. T   5. F   8. T
3. T   6. F   9. T

Ex. 5–2:

1. 28   4. 32   7. 5   10. 0
2. 36   5. 4   8. 54
3. 17   6. 19   9. 1

11. $(30 - 12) \div (3 \times 2)$
12. $[30 - (12 \div 3)] \times 2$
13. $[(30 - 12) \div 3] \times 2$
14. $30 - [12 \div (3 \times 2)]$
15. $30 - [(12 \div 3) \times 2]$

Ex. 5–3:

1. T   3. F   5. F
2. T   4. T   6. F

Ex. 5–4:

1. O   3. O   5. O
2. T   4. F   6. F

Ex. 5–5:

1. {2}
2. {—3}
3. {—3,—4,—5}
4. Ø
5. {4}
6. {—5,—4,—3,—2,—1,0,1,2,3,4}
7. {0}
8. {—1}
9. {—2}
10. {—5}

Ex. 5–6:

1. 7   3. 9   5. 4   7. 7
2. 7   4. —6 6. —5   8. —4

Ex. 5–7:

1. 14   3. 55   5. 77   7. 11   9. —22
2. 7   4. —17 6. —24 8. 8   10. 0

Ex. 5–8:

1. 40   4. 0   7. —25   10. 160
2. —27   5. 28   8. —9   11. —90
3. 210   6. —72 9. 56   12. —1

Ex. 5–9:

1. 9   4. 15   7. 52   10. 0
2. —7   5. —47 8. 13
3. —9   6. 9   9. 6

Ex. 5–10:

1. 5   5. 8   9. 10   13. 12
2. 40   6. —4   10. 73   14. 13
3. 9   7. —3   11. 9   15. 2
4. —5   8. 12   12. 2   16. 46

Ex. 5–11:

1. 2   4. 13   7. 6   10. 1
2. 2   5. 7   8. —12
3. 2   6. 2   9. 13

Ex. 5–12:

1. $n + 7$ 6. $2n + 9$
2. $2n - 3$ 7. $3n + 4n$
3. $n + 2n$ 8. $2n + 5$
4. $n + 8$ 9. $4n$
5. $5n + 10(7 - n)$ 10. $n - 3$

Ex. 5–13:

1. $t - 3 = 4$ 4. $t + 7 = 51$
2. $12t = 32$ 5. $t \div 5 = -9$
3. $3t - 8 = 7$

Ex. 5–14:

Typical solutions.

1. $2n + 6 = 22$; 8
2. $n + (n + 3) = 67$; 32 and 35
3. $r + (r - 9) = 239$; Roger 124 pounds and Ted 115 pounds
4. $p + (p + 8) = 26$; 9 feet
5. $2(w + 12) = 44$; 10 inches
6. $n + 4n = 75$; 15
7. $2a + 40 = 330$; \$1.45
8. $p(8 - 5) = 270$; 90 papers
9. $5n - n = 32$ or $n - 5n = 32$; 8 or —8

Ex. 6–1:

1. $\dfrac{17}{4}$　　4. 4　　7. 6　　10. −8

2. $-\dfrac{11}{3}$　　5. $\dfrac{20}{3}$　　8. −21

3. $\dfrac{9}{8}$　　6. $\dfrac{5}{2}$　　9. −9

Ex. 6–2:

1. $1\dfrac{3}{5}$　　4. $8\dfrac{6}{7}$　　7. $12\dfrac{4}{11}$

2. $4\dfrac{1}{10}$　　5. $4\dfrac{2}{9}$　　8. $9\dfrac{11}{25}$

3. $7\dfrac{2}{7}$　　6. $9\dfrac{4}{9}$　　9. $8\dfrac{21}{50}$

Ex. 6–3:

1. $\dfrac{1}{12}$　　4. $\dfrac{1}{40}$　　7. $\dfrac{1}{60}$

2. $\dfrac{1}{12}$　　5. $\dfrac{1}{18}$　　8. $\dfrac{1}{39}$

3. $\dfrac{1}{21}$　　6. $\dfrac{1}{30}$　　9. $\dfrac{1}{68}$

Ex. 6–4:

1. $\dfrac{12}{35}$　　5. $\dfrac{20}{39}$　　9. $\dfrac{15}{77}$

2. $\dfrac{10}{21}$　　6. $\dfrac{8}{45}$　　10. $\dfrac{15}{88}$

3. $\dfrac{40}{63}$　　7. $\dfrac{16}{39}$　　11. $\dfrac{10}{45}$ or $\dfrac{2}{9}$

4. 0　　8. $\dfrac{9}{16}$　　12. $\dfrac{21}{88}$

Ex. 6–5:

1. $\dfrac{1}{4}$　　3. $\dfrac{25}{28}$　　5. $\dfrac{5}{12}$　　7. $\dfrac{5}{9}$

2. $\dfrac{2}{5}$　　4. 12　　6. $\dfrac{3}{10}$　　8. $\dfrac{1}{8}$

Ex. 6–6:

1. $3\dfrac{3}{5}$　　3. $\dfrac{2}{5}$　　5. 12　　7. 112

2. $8\dfrac{1}{2}$　　4. 52　　6. 6　　8. $2\dfrac{1}{4}$

Ex. 6–7:

1. $\dfrac{8}{5}$　　4. $\dfrac{12}{7}$　　7. $\dfrac{2}{7}$

2. $\dfrac{1}{7}$　　5. $\dfrac{31}{9}$　　8. $-\dfrac{1}{9}$

3. $-\dfrac{9}{4}$　　6. $-\dfrac{37}{15}$　　9. $-\dfrac{7}{47}$

Ex. 6–8:

1. 2　　6. $2\dfrac{3}{11}$　　11. $6\dfrac{3}{10}$

2. $1\dfrac{1}{2}$　　7. $4\dfrac{1}{2}$　　12. $\dfrac{1}{20}$

3. $\dfrac{5}{24}$　　8. $10\dfrac{1}{2}$　　13. 5

4. $1\dfrac{5}{11}$　　9. $1\dfrac{2}{3}$　　14. $\dfrac{2}{3}$

5. $1\dfrac{7}{9}$　　10. $6\dfrac{2}{5}$　　15. $1\dfrac{1}{3}$

Ex. 6–9:

1. 1　　5. $\dfrac{2}{3}$　　9. 3

2. 1　　6. $\dfrac{3}{4}$　　10. $7\dfrac{2}{3}$

3. 1　　7. $\dfrac{2}{3}$　　11. $1\dfrac{5}{7}$

4. $\dfrac{7}{13}$　　8. $\dfrac{3}{5}$　　12. $7\dfrac{3}{5}$

Ex. 6–10:

1. $\dfrac{17}{21}$ 5. $\dfrac{41}{63}$ 9. $\dfrac{53}{72}$

2. $\dfrac{5}{9}$ 6. $\dfrac{1}{3}$ 10. $\dfrac{59}{60}$

3. $\dfrac{9}{10}$ 7. $\dfrac{35}{36}$ 11. $\dfrac{79}{120}$

4. $1\dfrac{5}{12}$ 8. $\dfrac{43}{77}$ 12. $1\dfrac{1}{12}$

Ex. 6–11:

1. $11\dfrac{1}{2}$ 4. $72\dfrac{7}{10}$ 7. $297\dfrac{11}{12}$

2. $26\dfrac{1}{4}$ 5. $34\dfrac{9}{40}$ 8. $806\dfrac{13}{14}$

3. $21\dfrac{5}{12}$ 6. $75\dfrac{5}{8}$ 9. $801\dfrac{49}{60}$

Ex. 6–12:

1. $\dfrac{4}{9}$ 6. $\dfrac{11}{35}$ 11. $2\dfrac{5}{12}$

2. $\dfrac{1}{3}$ 7. $\dfrac{1}{9}$ 12. $2\dfrac{1}{40}$

3. $4\dfrac{2}{5}$ 8. $\dfrac{2}{25}$ 13. $7\dfrac{13}{18}$

4. $3\dfrac{1}{10}$ 9. $\dfrac{13}{48}$ 14. $30\dfrac{39}{56}$

5. $\dfrac{2}{33}$ 10. $\dfrac{3}{40}$ 15. $10\dfrac{38}{45}$

Ex. 6–13:

1. .47   3. .68   5. .112
2. .8   4. .75   6. .65

7. $12\dfrac{7}{10}$ 9. $\dfrac{1}{8}$ 11. $\dfrac{17}{500}$

8. $\dfrac{37}{10000}$ 10. $\dfrac{3}{5}$ 12. $\dfrac{19}{400}$

Ex. 6–14:

1. 55.21   3. 284.3   5. 12.99069
2. 11.375   4. 1.5222   6. 1.222291

7. 43.6   9. 53.89   11. 3.8957
8. 169.9   10. .039   12. .15057

Ex. 6–15:

1. 2.72   3. 202.32   5. 25.9386
2. 3.8   4. 1.8944   6. 67.854

Ex. 6–16:

1. 18.5   3. 54.3
2. .28   4. 14.7

Ex. 6–17:

1. .35   3. .275   5. 2.75
2. .875   4. .38   6. 58.36

Ex. 6–18:

1. .83$\overline{3}$ 3. .6$\overline{6}$ 5. .7$\overline{857142}$
2. .1$\overline{1}$ 4. .15$\overline{15}$ 6. .$\overline{571428}$

Ex. 6–19:

1. $\dfrac{35}{99}$ 3. $\dfrac{44}{111}$ 5. $\dfrac{22}{30}$

2. $5\dfrac{2}{3}$ 4. $\dfrac{2105}{3333}$ 6. $2\dfrac{431}{1650}$

Ex. 6–20:

1. Yes   3. Yes   5. No
2. No   4. No   6. No

Ex. 7–1:

1.–4. Construction
5. No 6. Yes 7. No 8. No 9. Yes
10. No

Ex. 7–2:

1. Q   3. B   5. $\overline{TQ}$ 7. S
2. $\overline{RS}$ 4. Ø   6. $\overrightarrow{QS}$ 8. T

Ex. 7–3:

1. f   4. e   7. h   10. i
2. g   5. b   8. j
3. c   6. d   9. a

Ex. 7–4:

1. 6   3. Triangle 5. 5
2. No, two interiors 4. 2

Ex. 7–5:

1. Common side not between them
2. V   4. *d*
3. S or R   5. *a* or *c*
6. No, point P on the angle

7. $\overrightarrow{PT}$ and $\overrightarrow{PR}$

Ex. 7–6:

1. 45° 2. 125° 3. 90° 4. 180°

Ex. 7–7:

1. Straight; obtuse; acute
2. 119°; 61°; 180°
3. $\overleftrightarrow{MN}$

Ex. 7–8:

1. Indefinite number 4. Yes
2. Only one   5. Not necessarily
3. Yes

Ex. 8–1:

1. 2 in., $2\frac{1}{2}$ in.   3. 3 in., $2\frac{1}{2}$ in.

2. 2 in., $2\frac{0}{2}$ in.   4. 1 in., $1\frac{0}{2}$ in.

Ex. 8–2:

1. $\frac{1}{2}$ in.; $\frac{1}{4}$ in.; $\left(14\frac{1}{2}\pm\frac{1}{4}\right)$ in.

2. $\frac{1}{4}$ ft.; $\frac{1}{8}$ ft.; $\left(3\frac{1}{4}\pm\frac{1}{8}\right)$ ft.

3. $\frac{1}{8}$ mi.; $\frac{1}{16}$ mi.; $\left(5\frac{7}{8}\pm\frac{1}{16}\right)$ mi.

4. 1 in.; $\frac{1}{2}$ in.; $\left(9\pm\frac{1}{2}\right)$ in.

5. $\frac{1}{6}$ yd.; $\frac{1}{12}$ yd.; $\left(57\frac{0}{6}\pm\frac{1}{12}\right)$ yd.

6. $\frac{1}{16}$ in.; $\frac{1}{32}$ in.; $\left(15\frac{9}{16}\pm\frac{1}{32}\right)$ in.

Ex. 8–3:

1. .1 ft.; .05 ft.
2. .01 in.; .005 in.
3. .001 in.; .0005 in.
4. .01 ft.; .005 ft.
5. 100 ft.; 50 ft.
6. 10 ft.; 5 ft.

Ex. 8–4:

1. 10 ft.; 5 ft.; 16.7%
2. .1 in.; .05 in.; 6.7%
3. .01 mi.; .005 mi.; .4%
4. .001 in.; .0005 in.; 0%
5. 100 mi.; 50 mi.; 7.1%
6. 10 yd.; 5 yd.; .3%

Ex. 8–5:

1. 90.4   3. 3.19 5. 45000
2. 10.492 4. 3.62 6. 67.2

Ex. 8–6:

1. 4 4. 2 7. 2
2. 4 5. 3 8. 3
3. 4 6. 3 9. 5

Ex. 8–7:

1. 220   4. 34000   7. .042
2. .08   5. 3.91   8. 200
3. 87.1 6. .71   9. 160̲
                      10. 62̲

Ex. 8–8:

1. 600   4. 7.8   7. 100,000
2. 8.34 5. 9000   8. 876
3. 5700 6. 32   9. 350
                    10. 3.58

Ex. 8–9:

1. 30.48 3. 8.19 5. .303   7. 21.39
2. 6.37   4. 16.5 6. 30.48 8. 6.86

Ex. 8–10:

1. 5000 3. 8000 5. 2,700,000
2. 2.73 4. 3.85 6. 500

Ex. 8–11:

1. 5000 3. 49   5. 24.7
2. 2.8   4. .28 6. 33.9

Ex. 8–12:
1. 392 3. 167 5. 239
2. 35 4. 55 6. 20

Ex. 9–1:
1. 525 4. 350
2. 675 5. 240
3. 1000

Ex. 9–2:
1. $\dfrac{2}{3}$ 3. $\dfrac{1}{3}$ 5. $\dfrac{3}{2}$ 7. $\dfrac{12}{13}$

2. $\dfrac{9}{2}$ 4. $\dfrac{4}{11}$ 6. $\dfrac{35}{3}$ 8. $\dfrac{5}{6}$

Ex. 9–3:
1. $\dfrac{1}{8}$ 4. $\dfrac{7}{60}$ 7. 5° in 1 hr.

2. $\dfrac{24}{1}$ 5. $\dfrac{1}{4}$ 8. $15 for 1 ton

3. $\dfrac{10}{1}$ 6. $\dfrac{2}{3}$ 9. $15 in 1 hr.

10. 50 gal. in 1 min.
11. 300 mi. in 1 hr.
12. 100 ft. in 1 sec.

Ex. 9–4:
1. $\dfrac{8}{20} = \dfrac{12}{n}$; $n = 30$

2. $\dfrac{2}{3} = \dfrac{3}{n}$; $n = 4\dfrac{1}{2}$

3. $\dfrac{320}{5} = \dfrac{n}{7}$; $n = 448$

4. $\dfrac{1}{12 \times 18} = \dfrac{1}{18 \times n}$; $n = 12$

5. $\dfrac{72}{5} = \dfrac{n}{22}$; $n = 316.8$

6. $\dfrac{3}{7} = \dfrac{n}{126{,}000}$; $n = 54{,}000$

7. $\dfrac{1}{75} = \dfrac{6}{n}$; $n = 450$

8. $\dfrac{140}{560} = \dfrac{n}{8000}$; $n = 2000$

Ex. 9–5:
1. $\dfrac{3}{4} = \dfrac{5}{n}$; $n = 6\dfrac{2}{3}$

2. $\dfrac{300}{135} = \dfrac{1000}{n}$; $n = 450$

3. $\dfrac{55}{25} = \dfrac{3465}{n}$; $n = 1575$

4. $\dfrac{3}{5} = \dfrac{n}{65}$; $n = 39$

Ex. 9–6:
1. 36% 5. 85% 9. 40%
2. 52% 6. 80% 10. 30%
3. 70% 7. 45% 11. 68%
4. 15% 8. 5% 12. 85%

13. .47 15. .95 17. 1.5
14. 1 16. .13 18. .006

19. 2 20. 3.5 21. 10

Ex. 9–7:
1. 20% 3. 80 5. 25%
2. 10.08 4. 52.9 6. 197.75

Ex. 9–8:
1. $130 3. 12 5. 75%
2. 15 4. 392

Ex. 9–9:
1. 6% 3. $63 5. $48
2. $1125 4. 4%

Ex. 10–1:
1. Yes 2. $5\dfrac{1}{2}$ 3. Yes 4. Yes

Ex. 10–2:
1.–4. Constructions

Ex. 10–3:
1.–2. Constructions

Ex. 10–4:
1.–2. Constructions

Ex. 10–5:
1. $\triangle$ABC $\cong$ $\triangle$EDF
2. $\triangle$PQR $\cong$ $\triangle$LKJ
3. $\triangle$GEF $\cong$ $\triangle$SRT
4. $\triangle$TUV $\cong$ $\triangle$ZYX

Ex. 10–6:
1. A.S.A. 2. S.S.S. 3. S.A.S.

Ex. 10–7:

1. $\angle CAB \cong \angle DAB$ 3. $\overline{SR} \cong \overline{QP}$

$\angle CBA \cong \angle DBA$ $\overline{PR} \cong \overline{PR}$

$\overline{AB} \cong \overline{AB}$ $\angle SRP \cong \angle QPR$

$\triangle ABC \cong \triangle ADB$ $\triangle PRS \cong \triangle RPQ$

     A.S.A.   S.A.S.

2. $\overline{ED} \cong \overline{FD}$ 4. $\overline{MN} \cong \overline{KN}$

$\overline{EG} \cong \overline{FG}$ $\overline{JN} \cong \overline{LN}$

$\overline{DG} \cong \overline{DG}$ $\angle MNL \cong \angle KNJ$

$\triangle EDG \cong \triangle FDG$ $\triangle MNL \cong \triangle KNJ$

     S.S.S.   S.A.S.

Ex. 10–8:
1. g 4. 60° 7. 120°
2. d 5. 60° 8. 120°
3. c 6. 120° 9. 60°

Ex. 10–9:

1. $\overrightarrow{CD}$ bisects $\angle ACB$ Given
   $\angle ACD \cong \angle BCD$ Def. of bisect

   $\overline{AC} \cong \overline{BC}$ Given

   $\overline{CD} \cong CD$ Identity congruence
   $\triangle ADC \cong \triangle BDC$ S.A.S

2. $\overleftrightarrow{ML} \parallel \overleftrightarrow{JK}$ Given
   $\angle MLJ \cong \angle KJL$ Alt. int. angles $\cong$

   $\overleftrightarrow{MJ} \parallel \overleftrightarrow{LK}$ Given
   $\angle MJL \cong \angle KLJ$ Alt. int. angles $\cong$

   $\overline{JL} \cong \overline{JL}$ Identity congruence
   $\triangle JKL \cong \triangle LMJ$ A.S.A.

3. $\overline{SR}$ bisects $\overline{PQ}$ Given

   $\overline{PT} \cong \overline{QT}$ Def. of bisect

   $\overline{PQ}$ bisects $\overline{SR}$ Given

   $\overline{ST} \cong \overline{RT}$ Def. of bisect
   $\angle PTS \cong \angle QTR$ Vert. angles $\cong$
   $\triangle PTS \cong \triangle QTR$ S.A.S.

Ex. 10–10:
1. 40 2. 70 3. 4 4. 6

5. $\overline{DE} \parallel \overline{AB}$ Given
    $\angle A \cong \angle CDE$ Corr. angles $\cong$
    $\angle B \cong \angle CED$ Corr. angles $\cong$
    $\angle C \cong \angle C$ Identity congruence
    $\triangle ABC \cong \triangle DEC$ Angles of one $\triangle \cong$
    respectively to the
    angles of the other $\triangle$

Ex. 10–11:
1. 90 2. 50 3. 60

Ex. 10–12:

1. $\dfrac{x}{42} = \dfrac{8}{12}$; $x = 28$; 28 ft.

2. $\dfrac{10}{12} = \dfrac{30}{x}$; $x = 36$; 36 yd.

3. $\dfrac{x}{24} = \dfrac{6}{8}$; $x = 18$; 18 ft.

Ex. 11–1:
1. 60 ft.   3. 19 in.   5. 23.84 in.
2. 172 in.   4. 22 yd.   6. 1966 ft.

Ex. 11–2:
1. 36 in.   3. 7.2 ft.   5. 14 yd.
2. 60 ft.   4. 52 in.   6. 5.04 in.

7. 7 in.   9. 29 in.   11. 2.3 ft.
8. 20 ft.   10. 18 yd. 12. 1.24 in.

Ex. 11–3:
1. 6 in. 2. 4.3 ft. 3. Indefinite number

4. Indefinite number

5. 10.048 or 10 in.   8. 27.004 or 27 in.
6. 18.84 or 19 ft.   9. 75.36 or 75 ft.
7. 75.36 or 75 in.   10. 94.2 or 94 yd.

Ex. 11–4:
1. 29.42 or 29 in.   3. 31 in.
2. 23.55 or 24 in.   4. 12.56 or 13 in.

Ex. 11–5:
1. 16 or 20 sq. ft.
2. 132 or 130 sq. in.
3. 198 or 200 sq. yd.
4. 27.2 or 27 sq. in.
5. .3225 or .32 sq. ft.
6. 26.3088 or 26.3 sq. in.

Ex. 11–6:
1. 25 or 20 sq. in.   4. 18.49 or 18 sq. in.
2. 64 or 60 sq. in.   5. 132.25 or 132 sq. ft.
3. 289 or 290 sq. ft.   6. 441 or 440 sq. yd.

Ex. 11–7:
1. 56   3. 23.37
2. 150   4. 756

Ex. 11–8:
1. 90   3. 20.44
2. 338   4. 210

Ex. 11–9:
1. 12.56 sq. in.   5. 1962.5 sq. in.
2. 50.24 sq. ft.   6. 1098.0266 sq. ft.
3. 452.16 sq. yd.   7. 50.24; 200.96 sq. ft.
4. 32.1536 sq. in.   8. 15,386

Ex. 11–10:
1. 280 or 300 cu. in.
2. 2496 or 2000 cu. ft.
3. 32.256 or 32 cu. in.
4. 378 or 400 cu. yd.
5. 17.5 or 20 cu. ft.
6. 11,700 or 10,000 cu. ft.
7. 36 or 40 cu. ft.
8. 2250 or 2000 pounds

Ex. 11–11:
1. 549.5 or 500 cu. in.
2. 37.68 or 40 cu. ft.
3. 33.761 or 34 cu. in.
4. 2204.28 or 2000 cu. in.
5. 15.7 or 20 cu. yd.
6. 169.56 or 200 cu. in.
7. 235.5 or 200 cu. in.

Ex. 12–1:
1. $\dfrac{1}{11}$ 2. $\dfrac{6}{11}$ 3. $\dfrac{6}{11}$ 4. $\dfrac{2}{11}$ 5. $\dfrac{5}{11}$
6. 0 7. $\dfrac{1}{7}$ 8. $\dfrac{5}{14}$ 9. $\dfrac{1}{2}$ 10. $\dfrac{6}{7}$
11. $\dfrac{9}{14}$ 12. $\dfrac{1}{2}$ 13. $\dfrac{5}{14}$ 14. $\dfrac{1}{2}$

Ex. 12–2:
1. $\dfrac{1}{3}$ 2. $\dfrac{2}{3}$ 3. $\dfrac{4}{9}$ 4. $\dfrac{5}{9}$ 5. $\dfrac{2}{9}$
6. $\dfrac{7}{9}$ 7. $\dfrac{1}{26}$ 8. $\dfrac{25}{26}$ 9. $\dfrac{5}{26}$ 10. $\dfrac{21}{26}$
11. $\dfrac{2}{13}$ 12. $\dfrac{11}{13}$ 13. $\dfrac{7}{26}$ 14. $\dfrac{19}{26}$

Ex. 12–3:
1. $\dfrac{1}{13}$ 2. $\dfrac{4}{13}$ 3. $\dfrac{3}{13}$ 4. $\dfrac{3}{13}$ 5. $\dfrac{7}{26}$

Ex. 12–4:
1. $\dfrac{9}{100}$ 4. $\dfrac{16}{81}$ 7. $\dfrac{2}{27}$ 10. $\dfrac{16}{81}$
2. $\dfrac{49}{100}$ 5. $\dfrac{8}{81}$ 8. $\dfrac{4}{27}$
3. $\dfrac{21}{100}$ 6. $\dfrac{4}{81}$ 9. $\dfrac{1}{9}$

Ex. 12–5:
1. $\dfrac{1}{15}$ 4. $\dfrac{5}{33}$ 7. $\dfrac{35}{132}$ 10. $\dfrac{4}{663}$
2. $\dfrac{7}{15}$ 5. $\dfrac{7}{22}$ 8. $\dfrac{1}{221}$ 11. $\dfrac{8}{663}$
3. $\dfrac{7}{30}$ 6. $\dfrac{35}{132}$ 9. $\dfrac{13}{204}$

Ex. 12–6:

1. 32   3. $\dfrac{5}{16}$ 5. 1

2. $\dfrac{1}{32}$ 4. $\dfrac{5}{32}$

6. 1, 6, 15, 20, 15, 6, 1
7. 1, 7, 21, 35, 35, 21, 7, 1

Ex. 12–7:

1. 5   2. 90   3. 8   4. 10

Ex. 12–8:
1. 85   2. 12   3. No mode

Ex. 12–9:
1. 67   2. 136   3. 86

4. Brand D, mode
5. Mode 30¢, Median 34¢, Mean 39¢; use either median or mode.
6. Mean 53⅛, Median 50; yes. Modes 12 and 105; no.

GLOSSARY

ACUTE ANGLE. An angle whose measure is greater than 0° but less than 90°.

ALTITUDE (of a triangle). The perpendicular line segment joining any vertex of the triangle to the line that contains the opposite side of the triangle.

ANGLE. The union of two rays that have the same end point.

ARC. Any part of a circle containing more than one point.

AREA. The measurement of the interior of a closed figure.

ASSOCIATIVE PROPERTY OF ADDITION. For all numbers a, b, and c, $(a+b)+c = a+(b+c)$.

ASSOCIATIVE PROPERTY OF MULTIPLICATION. For all numbers a, b, and c, $(a \times b) \times c = a \times (b \times c)$.

CIRCLE. The set of all points in a plane at a given distance from some point in the plane.

CIRCUMFERENCE. The perimeter of a circle.

CLOSURE. If a and b are any two numbers of set A and * denotes an operation, then set A is closed under (or with respect to) the operation * if $a*b$ is a member of set A.

COMMUTATIVE PROPERTY OF ADDITION. For all numbers a and b, $a+b=b+a$.

COMMUTATIVE PROPERTY OF MULTIPLICATION. For all numbers a and b, $a \times b = b \times a$.

CONGRUENT ANGLES. Two angles that have the same measure.

CONGRUENT LINE SEGMENTS. Two line segments which have the same measure.

CONGRUENT TRIANGLES. Two triangles that have exactly the same size and the same shape, or two triangles whose corresponding parts are congruent.

DENOMINATOR. The integer b in the rational number $\frac{a}{b}$.

DENSITY PROPERTY (of rational numbers). Between any two rational numbers there is a third rational number.

DIAMETER. A line segment that joins two points of a circle and passes through the center of the circle.

DISJOINT SETS. Sets that have no members in common.

DISTRIBUTIVE PROPERTY. For all numbers a, b, and c, $a \times (b+c) = (a \times b) + (a \times c)$.

EMPTY SET. The set that contains no members.

EQUATIONS. Those number sentences that state that two expressions are names for the same number.

EQUIVALENT SETS. Sets between which there is a one-to-one correspondence.

FRACTION. A name for a rational number.

IDENTITY CONGRUENCE. A congruence between a geometric figure and itself.

INTEGERS. $\{ \ldots, -3, -2, -1, 0, 1, 2, 3, \ldots \}$

INTEREST. The money one pays for the use of money borrowed.

INTERSECTION OF SETS. The intersection of set A and set B, denoted by $A \cap B$, is the set of all objects that are members of both set A and set B.

INVERSE OPERATION. Any process or operation that "undoes" another process or operation.

IRRATIONAL NUMBERS. The set of numbers each of which can be named by a nonterminating and nonrepeating decimal.

LINE SEGMENT. Line segment AB is the union of points A and B and all the points on line AB that are between points A and B.

MEAN (ARITHMETIC MEAN). The average of the numbers in a set.

MEDIAN. The number that is in the middle position when numbers are arranged in their natural order, either least to greatest or greatest to least.

MEMBERS. The things contained in a given set.

MODE. The number that occurs most frequently in a set of numbers.

MULTIPLICATIVE INVERSE. See Reciprocal.

NATURAL NUMBERS. The set of numbers 1, 2, 3, 4, 5, . . . Also called the set of counting numbers.

NUMERAL. A name for a number.

NUMERATION SYSTEM. A planned scheme or way of naming numbers.

NUMERATOR. The integer a in the rational number $\dfrac{a}{b}$.

OBTUSE ANGLE. An angle whose measure is greater than 90° but less than 180°.

ONE-TO-ONE CORRESPONDENCE. The relationship between two sets when every member of each set is paired with one and only one member of the other set.

OPEN SENTENCES. Mathematical sentences that contain letters (or some other symbols) to to be replaced by numerals, and that are neither true nor false.

PARALLEL LINES. Two lines in a plane whose intersection is the empty set.

PARALLEL PLANES. Two planes whose intersection is the empty set.

PERIMETER. The distance around or the measurement of the boundary of a figure.

PERPENDICULAR LINES. Two lines that intersect so that right angles are formed.

PLACE VALUE. The value denoted by a digit because of its position in the simplest numeral for a number.

POLYGON. A simple closed figure formed by the union of line segments that have common end points.

PRINCIPAL. The amount of money one borrows or lends.

PROBABILITY. The ratio of the number of successful outcomes of an event to the number of possible outcomes of the event.

PROPORTION. The equality of two ratios.

RADIUS. A line segment joining the center of a circle with any point on the circle.

RATIO. The relationship, expressed as a fraction, between the numbers of two sets.

RATIONAL NUMBER. Any number of the form $\dfrac{a}{b}$ where a and b are integers and $b \neq 0$ (the quotient of two integers where the divisor is not zero).

RAY. Ray AB is the union of point A and all points on line AB that are on the same side of point A as point B.

REAL NUMBERS. The union of the set of rational numbers and the set of irrational numbers.

RECIPROCAL (MULTIPLICATIVE INVERSE). If the product of two rational numbers is 1, then the two rational numbers are reciprocals of each other.

RECTANGLE. A quadrilateral in which opposite sides are parallel and all angles are right angles.

REPLACEMENT SET. The set of numbers whose numerals are to be used as replacements for a variable.

RIGHT ANGLE. An angle whose measure is 90°.

ROOT. A solution of an open sentence, or any element of the solution set.

SET EQUALITY. If A and B are names for sets, $A = B$ means that set A has identically the same members as set B, or A and B are two names for the same set.

SIMILAR TRIANGLES. Two triangles whose corresponding angles are congruent and whose corresponding sides are proportional.

SIMPLE CLOSED FIGURES. Figures which separate a plane into three distinct sets of points—the figure itself, the interior, and the exterior of the figure.

SOLUTION SET. The set of replacements for the variables in an open sentence that make the resulting sentence true.

SOLVE AN EQUATION. To find the solution set of an equation.

SPACE. The set of all points.

SQUARE. A rectangle in which all four sides are congruent to each other.

STRAIGHT ANGLE. An angle whose measure is 180°.

SUBSET. Set A is a subset of set B if every member of set A is also a member of set B.

UNION OF SETS. The union of set A and set B, denoted by $A \cup B$, is the set of all objects that are members of set A, of set B, or of both set A and set B.

WHOLE NUMBERS. $\{0,1,2,3,4,5, \ldots\}$

INDEX